# INTRODUCING CICHLIDS

*Richard F. Stratton*

Distributed in the UNITED STATES to the Pet Trade by T.F.H. Publications, Inc., 1 TFH Plaza, Neptune City, NJ 07753; on the Internet at www.tfh.com; in CANADA by Rolf C. Hagen Inc., 3225 Sartelon St., Montreal, Quebec H4R 1E8; Pet Trade by H & L Pet Supplies Inc., 27 Kingston Crescent, Kitchener, Ontario N2B 2T6; in ENGLAND by T.F.H. Publications, PO Box 74, Havant PO9 5TT; in AUSTRALIA AND THE SOUTH PACIFIC by T.F.H. (Australia), Pty. Ltd., Box 149, Brookvale 2100 N.S.W., Australia; in NEW ZEALAND by Brooklands Aquarium Ltd., 5 McGiven Drive, New Plymouth, RD1 New Zealand; in SOUTH AFRICA by Rolf C. Hagen S.A. (PTY.) LTD., P.O. Box 201199, Durban North 4016, South Africa; in JAPAN by T.F.H. Publications. Published by T.F.H. Publications, Inc.

MANUFACTURED IN THE
UNITED STATES OF AMERICA
BY T.F.H. PUBLICATIONS, INC.

This jewel cichlid (*Hemichromis lifalili*) is one of those cichlids notorious for bad temper, especially among its own kind. Keeping jewel cichlids is well worth the effort, however, as they are fabulously handsome fish that practice intense parental care. Photo by H. J. Richter.

# CONTENTS

THE APPEAL OF CICHLIDS .......... 4

THE CICHLID TANK .......... 12

THE KINDS OF CICHLIDS .......... 21

THE CICHLIDS OF CENTRAL AND SOUTH AMERICA .......... 25

AFRICAN CICHLIDS .......... 47

# THE APPEAL OF CICHLIDS

The popularity of cichlids among advanced aquarists is astounding. It was not always this way. There was a time when only the fanatical kept cichlids. That was in the days of small aquaria and small fishes and planted tanks. Cichlids either killed or ate the different fish species and uprooted the plants. It is not surprising, then, that they were unpopular among general fish keepers. However, there were a select few who were willing to go the extra mile, to make the extra effort that was required in keeping cichlids. That was because they saw something special in them. But why did they become so popular? How did so many hobbyists discover the special qualities of cichlids?

Part of the reason for the increase in cichlid popularity was simply that bigger tanks became possible with the development of sealants that were used in the space program. With larger tanks available at reasonable prices, people could keep larger fish species, including cichlids. Cichlids particularly appealed to people because of their human-like behavior. That is, they had purposeful movements throughout the tank, easily discerned by humans. A cichlid will swim across the tank to look in a cave to inspect it for a possible nest site or just out of seeming curiosity.

It is probably a human trait to identify with animals that seem like us. If primates weren't such nuisances to keep, we might keep them instead of dogs and cats. But we can even see the "human" traits in dogs and cats. And cichlids afford that, too. One reason is that they take care of their young, caring for them with a tender solicitousness and defending them with a fierce vigor. In fact, that is why many cichlid species would kill other fish species, not out of viciousness but simply to defend their young or the territory in which eggs might be laid.

Cichlids even resemble us in the number of nostrils that they have. As strange as it

***Apistogramma agassizii*, Agassiz's dwarf cichlid, is a most desirable dwarf cichlid from the Amazon. Peaceful (except at spawning time) and not *too* delicate, this is a perfect fish for a small, well-planted tank. Photo by A. Norman.**

*Herichthys (Thorichthys) meeki*, commonly known as the firemouth. This Central American cichlid is a dynamite aquarium resident with its vivid reds and iridescence. Firemouths are not particularly troublesome—unless, of course, you mess with their family! Photo by H. J. Richter.

may seem, nearly all other fish species have four nostrils. The fish swims forward, and the water flows in one nostril and exits through the posterior one, with a set of nostrils on each side of the snout. The cichlids have simplified this structure, while at the same time improving upon it. A fish's nostrils are not connected with its respiratory system; hence, the flow-through system utilized by most fish species. However, such a system dictates that a fish must be in motion or there must be some current in order for the nostrils to function properly. The cichlid's nostrils act like a syringe, taking in water by muscular action, sampling, and then expelling it. In that manner, a fish can still "smell" the environment without having to depend on currents or movement of its own body.

Perhaps another appeal of cichlids is that people also prefer to identify with winners. Certainly they do in regard to sporting events. Everybody jumps on the bandwagon of a winning team, but only the faithful few suffer through the losing seasons. Are cichlids winners? They do have a propensity for wresting the control of a tank away from other fish species. (That is also one of their faults!) But the family Cichlidae (sick-lid-dee) is certainly a winner in the biological sense. In fact, cichlids are the biggest success story of all vertebrates. They have conquered more different types of habitats than other vertebrates and developed into more specialized different species, and only cichlid species (only a few of them though) are able to spawn in either ocean water or fresh water. In any case, it is our nature to love a winner, and certainly cichlids are a grand success story.

In the late 1960s, there began an importation of the cichlids of the Great Lakes of Africa, principally from Lake Malawi (then called "Nyassa") and Lake Tanganyika. We always knew that those species were there, but most of us only had the line drawings from Boulenger's *Catalogue of the Fishes of Africa*.

**The Texas cichlid, *Herichthys cyanoguttatus*, is no silly little fish. This bruiser will make short work of any tankmate that doesn't measure up. The specimen shown is a young male. Photo by A. Norman.**

# THE APPEAL OF CICHLIDS

*Aulonocara hansbaenschi*, one of the highly desirable Malawi peacock cichlids. Photo by Ad Konings.

We knew of the bizarre shapes, but we didn't realize just how colorful so many of the species were. Many of the species of Lake Malawi were absolutely beautiful, and they had the further virtue of supplying lots of motion and color throughout the tank. The mbuna, or rock-dwelling cichlids, were quite active, as opposed to the sedate lurking behavior of many other cichlid species. Needless to say, the sight of such beautiful fish in tanks piqued the interest of the general public. Many more people began to keep cichlids. Since Lakes Malawi, Tanganyika, and Victoria are virtual cichlid fish bowls, it was then possible to have community tanks of only cichlids and still have lots of variety in those tanks. Thus, the problem of tankmate bullying was alleviated to some extent, although these cichlids do tend to pick on each other—but it is a more equal battle.

As the popularity of cichlids increased, it began to gather momentum. People who keep a certain type of fish tend to learn more about them, and the more people learned, the more reasons they discovered for appreciating cichlids. Ingenious indeed are the ways in which cichlids have evolved to care for their young. And also quite ingenious are the niches that cichlids have exploited for earning a living, but we will cover more about that later. For right now, let me just point out that there are herbivorous cichlids and there are predatory cichlids. This is the case with many other fish families, too, but cichlids have "seemingly created their own niches," as one excited researcher commented. As just one example, there are predators in both Lake Malawi and Lake Tanganyika that feign death, lie on their sides, turn a death-like mottled gray coloration, and then pounce upon the small fishes which come to feast upon the "corpse!" If this seems unusual and bizarre, just wait, you haven't heard anything yet. (We will cover adaptive behavior and feeding specializations in the species sections.)

Scientists caught on to cichlid study long before tropical fish hobbyists did. In fact, such researchers would report in their books or reports that cichlids were "popular with scientists but unpopular with aquarists." Some animals, such as the fruit fly and white rat, have been popular because of their ease of maintenance and reproduction. In this respect, cichlids have fallen short, as even the easiest cichlids present problems. Nevertheless, it should be mentioned that the hardy species are extremely so. I once had a friend who proclaimed that "you couldn't kill a cichlid with a hammer." And the fact is that many cichlid species are extremely hardy, resistant to disease, quite tolerant of water conditions, and adaptable

*Steatocranus casuarius.* This is an exceptional example of a male lionhead cichlid. Though he looks big and fierce, this fish is generally mild and only grows to about 4 inches. Photo by MP. & C. Piednoir, Aqua Press.

*Herichthys (Archocentrus) nigrofasciatus*, the famous convict cichlid, in the white color form, leading some of her nice school of newly free-swimming fry. Photo by R. Zukal.

to wide fluctuations in hardness and pH.

I would be remiss if I failed to mention that some cichlids are a challenge to keep and breed. Although there are aquarium strains of discus and angels that are quite adaptable and breed readily, the same species from the wild are almost a different animal; they have not descended from many generations of fish individuals that adapted to aquarium conditions and bred readily in the aquarium. Generally speaking, the cichlids of Lake Tanganyika can be almost as demanding as marine fish, as the water must be kept in good condition and the pH maintained. Even with the hardy Malawi cichlids, the pH must be kept up to a minimum value. Part of the problem is always finding the conditions that best fit the species, but certainly some species are much more delicate than others are. For those who want a challenge, it is certainly there among the species of the family Cichlidae, which is generally renowned for its hardy and aggressive species. Once again, cichlids have all bases covered, as there are plenty of easy-to-keep species available, along with those that are a challenge.

There are two major reasons that cichlids have caught the attention of scientists. One is that cichlids have extremely complex behavior in which both the male and female cooperate in the defense of eggs and young. This is highly unusual. Most fish don't protect the young, and the ones that do provide protracted care almost always involve only one parent, usually the male. Some sort of parental care is present throughout the cichlid family, and it may be one reason for the spectacular success of this group of fishes. In any case, such complex behaviors as cooperation between the parents and communication with the young give scientists a chance to study the details of how all of this is accomplished.

Aquarists had known about how discus feed their young off a specialized "milk" or body slime that both parents secrete. But it was scientists who discovered that this behavior was present in other species in a less obvious manner. This type of feeding probably evolved in discus because they tend to inhabit nutrient-poor areas in the Amazon drainage system where their fry would have poor foraging prospects, as microorganisms would be scarce. The feeding of the young in this species was quite obvious, as the young would cling to the sides of a parent, much like mammals nursing. With the other cichlids studied, some species secreted a special nourishing substance, but it was not an obligatory food for the fry, so they only occasionally made use of it. In fact, their feeding behavior was called "glancing" by the researchers who discovered it.

When tropical fish hobbyists finally discovered the family in a big way, there was much scientific literature for them to peruse if so inclined. In fact, the model for ethology (the study of animal behavior) is a book that is nearly always referenced in any article in the field, *An Introduction to the Study of the Ethology of Cichlid Fishes* by G. P. Baerends and J. M. Baerends-Van Roon. The book is highly recommended, as it sets the standard and gives the basics for observing animal behavior, and cichlids in particular.

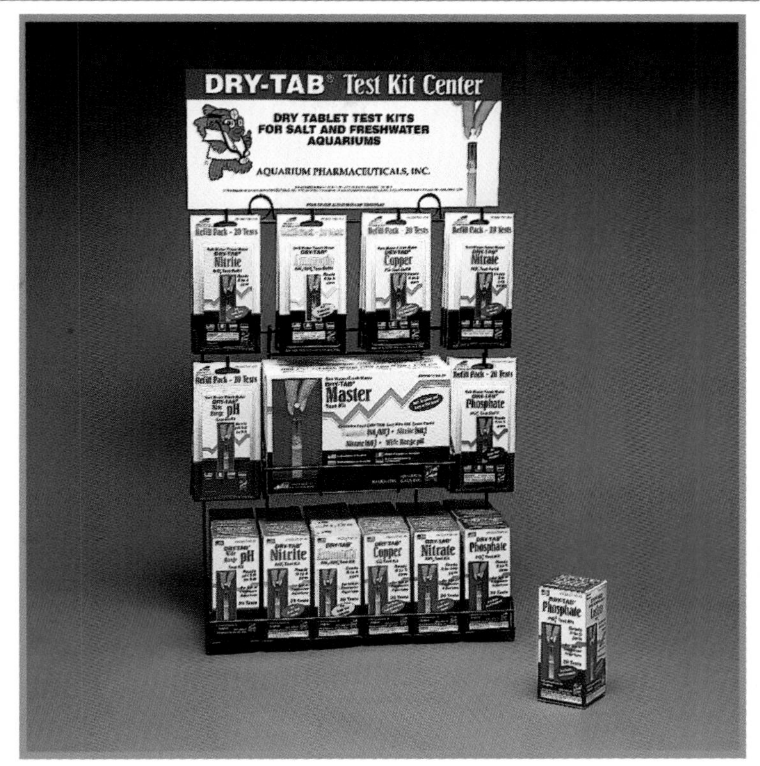


**The quickest way for hobbyists to check for unwanted chemicals in freshwater or saltwater aquariums is to test the water. The use of tablets that can simply be dropped into water samples, with resulting color changes matched against a color chart, makes testing easy. Photo courtesy of Aquarium Pharmaceuticals, Inc.**

Once the delights of the cichlid family were discovered, aquarists found that they could do without plants by compensating with interesting rockwork and driftwood. Also, with the family having become popular, the aquarium industry came up with ceramic artificial driftwood, caves, and rocks, which helped to beautify the tank and make the cichlids feel at home.

Part of the reason that cichlid popularity snowballed is that aquarists discovered that cichlids seemed to capture the prizes in all directions. Besides providing interesting behavior, the family has as many beautiful species as any other family, with both the king and queen of the aquarium world (the discus and the angel) being cichlids. If your taste runs more to the grotesque, there are some cichlids that are in the running for that category, too. The fact is that some species are even in the running for the most ugly fish. My first thought is *Steatocranus casuarius*, but I have a difficult time thinking of the species as ugly, as I am quite fond of it (as I am of nearly all cichlid species).

Some cichlids are even goofy-looking. My own nomination of such a species would be *Ophthalmotilapia nasuta* from Lake Tanganyika, although I am willing to concede that as a very subjective judgment. There are plenty of cichlidophiles (cichlid lovers) who find the species beautiful. Certainly it is interesting in behavior, as nearly all cichlids are. However, many are neither colorful nor exotic. They come in a plain brown wrapper and have very ordinary cichlid shapes. Even those species have their adherents, however, as some of them have quite interesting behavior.

Personality varies in cichlids, too, from the pugnacious red terrors of South America to the comical goby cichlids of Lake Tanganyika. The oscar has long been an aquarium favorite because of its "personality" and nearly dog-like behavior. One of the reasons it is so responsive to its keeper is that it is nearly always hungry and will practically jump through hoops for food. Since the owner is associated with food, the fish is always happy to see him or her and always keeps a close eye on the person. In any case, that is my observation on why oscars are so attentive to their owners and vice versa——but I am sure that many oscar owners feel that their fish have true affection for them, and I'm not about to dispute that, at least not out loud!

**This is a gold male *Ophthalmotilapia nasuta*. Photo by MP. & C. Piednoir.**

Experienced cichlid keepers will often keep a pet cichlid, too, but it is usually not an oscar. For fish with personality, it is more likely to be a red devil, a green terror, or a *Cichlasoma dovii*. Such specimens will learn to tolerate the owner but will often show extreme belligerence toward strangers. In such cases, the specimen will rail against the person's presence, hitting the glass with the snout, and when a fish such as a *C. dovii* does that, it sounds as though it is coming right through the glass! (In the meantime, the aquarist usually makes comments about the fish being a "good judge of character," or so it has been my experience!) For a beautiful single specimen, the candidate kept is usually an *Aulonocara* species from Lake Malawi, such as the Niagara peacock, or a red terror (*Herichthys festae*) from South America. I have also known *Boulengerochromis microlepis*, from Lake Tanganyika, to be kept as a single specimen as a pet. I was surprised that this open-water predator could adapt to the confined space of an aquarium. Obviously I was wrong, as the species has even been bred in captivity, and it is the largest of all cichlids, reaching three feet in length!

There are also tiny cichlids, barely attaining an inch in length. Many of these are in

***Boulengerochromis microlepis.*** **This is a big fish that will eat any other fish it can swallow. Photo by Mark Smith.**

the genus *Apistogramma* in South America, but some of the shell-dwelling species in Lake Tanganyika also weigh in at the smallest of sizes, and they are loaded with interesting and complicated behavior. I have often wondered how such intricate behavior could have been programmed into such a small cranial case, but perhaps that is what whales and elephants might think about us!

So diverse have the cichlids become in species and so widespread in habitats that some scientists have pondered breaking the family up into groups. Surely such specialized types as the discus and angels would fit into at least a subfamily of their own, it would seem. The same might be said about numerous specializations that have been taken up by the family, and some have suggested the myriad of African species must be separate families from the New World, or neotropical, species. Such has not proven to be the case. Analyses by scientists in recent years have indicated that cichlids of whatever species—as widespread and diverse as they are—still are monophyletic. That is, they are descended from a single marine species that invaded freshwater areas so long ago that the continents were in an entirely different position from where they are now. The ancestral marine form must have been quite a versatile species itself, as not many marine forms are able to compete against fish that have already evolved to fill freshwater niches. It is particularly impressive that cichlids were able to compete in the Amazon drainage areas, in which many specialized fishes have resided for millions of years and evolved to adapt to some of the specific conditions found in that area, such as the periodic flooding and expanding of the river. Freshwater fish species that are derived from marine ancestors are called "secondary freshwater fishes," as opposed to "primary freshwater fishes," which are the animals that have adapted from the start to freshwater habitats.

The fact is that some cichlids still invade the ocean and compete with the fish species there on a very successful basis. For tropical fish hobbyists, though, it is a felicitous fact that a marine ancestor gave rise to the cichlids, for they represent some of the most fascinating residents of the aquarium. With beauty, brawn, and brains in abundance, it is no wonder that cichlids have become such a hit in the aquarium world!

# THE CICHLID TANK

The cichlid aquarium will be different from other aquaria in several respects. First, depending on the cichlid species being kept, tankmates must be selected with care. Non-cichlid fishes should be of the type that inhabit the upper part of the water and are fast-swimming and agile. Of course, I am keeping in mind that people become so taken with cichlids that they put together community tanks composed only of cichlids, but I want to emphasize that any fish that are to be kept with cichlids should be selected with care.

## PLANTS

Plants, unfortunately, are going to be out in most cases, but you can compensate with colorful rockwork, driftwood, and plastic plants. Yes, I know that some hobbyists have succeeded in keeping some plants with cichlids, but these were people who really understood their fish species and the plants. I don't want to put anyone in the position of being unhappy with their cichlid because the fish just uprooted a valuable Amazon sword plant. Most cichlid species are not even plant eaters, but the plants take a lot of punishment, most particularly when cichlids are in the process of spawning. Nature has endowed most cichlid species with an abundance of aggression for the purpose of the protection of the eggs and of the young. The fish tear around the tank, looking for potential enemies of their young, and the plants get uprooted and ripped as the result of all this activity.

## MAINTAINING WATER QUALITY

Finally, a lot of thought should be given to filtration. Filtration is the life-support system of the aquarium and

**While plants may not feature importantly in your tank decor, there is no end to the arrangements you can make with rocks and driftwood. Be sure to leave an open area for swimming. If the tank is *too* crowded with decorations, it is difficult to clean the gravel as required. Photo by MP. & C. Piednoir.**

without some sort of filtration, your cichlids would not be long for this world. Because most cichlids are large, messy fish, you may elect to use several filters on your larger aquaria.

**Filtration**

For a lot of good reasons, the undergravel filter is one of the most popular of all filters. But it can be a problem with cichlids. That is because most cichlids tend to dig holes in the gravel. This is particularly true when spawning time arrives. In any case, such actions completely subvert the action of the undergravel filter. What happens is that the cichlids dig all the way down to the filter plate, exposing it. Since water tends to flow through the course of least resistance, all the fluid tends to flow through the area of the exposed filter plate, completely bypassing the gravel, which was intended as the filter medium. Hence, most cichlid keepers utilize some type of filtration other than an undergravel one.

It is possible, however, to use an undergravel filter with cichlids, but some modifications have to be made. A common solution is to lay down the gravel, then place in a piece of grating (such as that used as fluorescent light diffusers) on top of the gravel, and then add more gravel on top of that. Then when the fish try to dig, they can't get down to the filter plate. Other modifications include the use of plastic screening, which simply keeps the fish from digging below that level. Even modified, the undergravel filter is generally an unsatisfactory arrangement for cichlids. Most

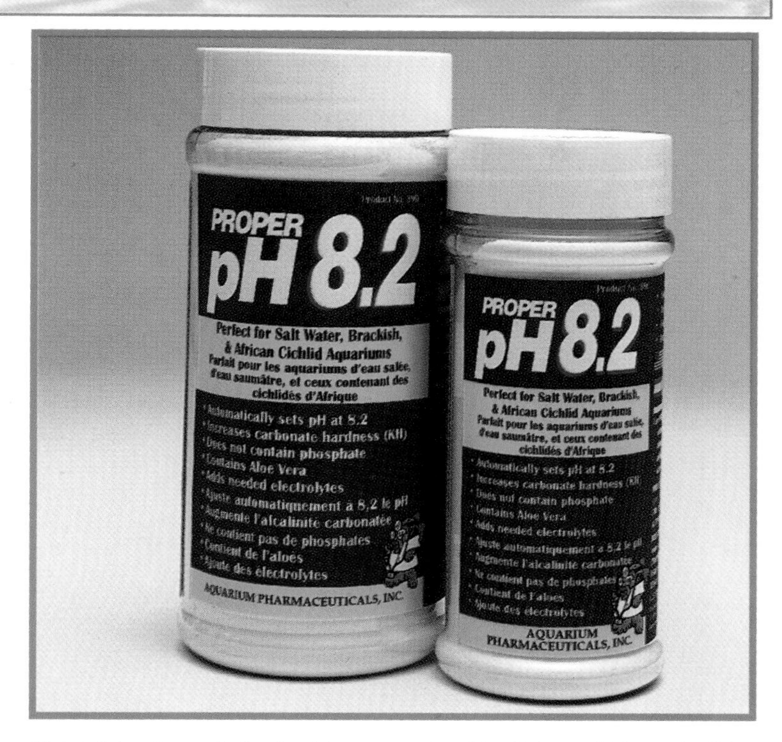


**Maintaining the water's pH at the proper level for fishes—African cichlids, marine, and brackish water species, for example—is a very important part of aquarium management. Products that allow hobbyists to achieve and maintain specified pH levels are inexpensive and easy to use. Photo courtesy of Aquarium Pharmaceuticals, Inc.**

**Canister filters can be used for mechanical, biological, and chemical filtration. They are very popular with cichlid-keepers who want to be able to use several types of media. Drawing by J.R. Quinn.**

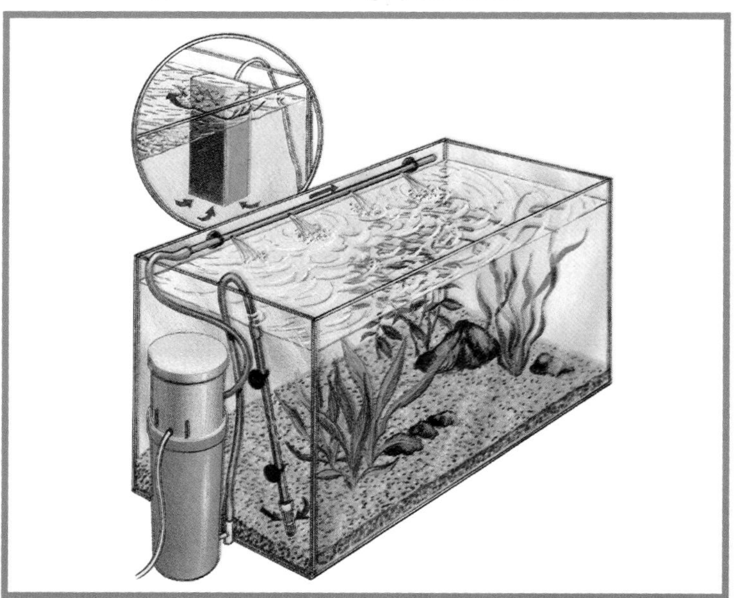

**A combination of rocks and driftwood can be used to create territories, spawning sites, and visual barriers that allows a greater variety of cichlids to live together in harmony. Photo by M.P. & C. Piednoir.**

cichlids are of a larger size than most other aquarium fish; hence, the gravel needs vacuuming more frequently, and it is difficult to get the gravel clean with grating or screening in the way.

One reason that the undergravel filter has been deservedly popular is that it takes care of two aspects of filtration, mechanical and biological. Mechanical filtration refers to the clearing of the water of debris. Chemical filtration refers to removing dissolved compounds by means of absorption and adsorption. Biological filtration refers to the breaking down of ammonia into less harmful compounds. Ammonia results from the decomposition of organic matter from the fish and from uneaten food. The ammonia is converted into nitrite. The nitrite is further reduced into nitrate. This is all done by bacterial activity. Ammonia and nitrite compounds are quite harmful to fish species, but nitrate can be tolerated in low levels. Unfortunately, the only bacteria that break up nitrate are anaerobic. That is, they function in the absence of oxygen. The unfortunate part is that toxic substances are produced by the reduction of the nitrate. For that reason, constructing a filter that breaks down nitrate has been difficult and fraught with hazards (to the fish).

Mechanical filtration usually involves the use of paper or plastic or sand or all three. Diatomaceous earth can also be used for fine filtration, which consists of the silica

**Retain the ability to rearrange objects when you are setting up your tank. There may come a day when you will have to rearrange territories to eliminate aggression. Photo by MP. & C. Piednoir.**

**Your fishes will show their best colors against dark gravel. The contrast is appealing and the fishes will be more confident when kept in a tank with a dark substrate. Photo by MP. & C. Piednoir.**

"skeletons" of tiny diatoms. Chemical filtration is usually done with activated carbons, but it can also involve ion exchange resins. Generally speaking, a filter gets added points if it has all three types of filtration. That is because the object of filtering is to remove debris from the water and also get rid of harmful substances that are dissolved in the water.

With the foregoing in mind, there are several filters that can be recommended for cichlid tanks. One of the old standbys is also the simplest. That is the inside box filter. It is powered by air, it runs the water through the medium quickly enough, and it provides all three types of filtration. The filter normally contains activated carbon for the chemical filtration and some sort of synthetic floss for the mechanical filtration. Eventually, the carbon and floss are colonized by the desirable bacteria so that biological filtration is also present, albeit weakly. A good strategy with box filters is to have several of them in a tank and clean a different set of them every two weeks. That way the bacteria will remain operative in the filters not yet cleaned and will have a chance to build up in the newly cleaned filters.

Part of the decision of what filter to use is determined by what the hobbyist wants the

**Undergravel filters provide excellent biological filtration. If you know the fishes you are going to put into the tank are diggers, simply cover the filter plate with a screen. Photo by I. Francais.**

tank to look like. In breeding tanks, inside filters are perfectly acceptable because no one is going to make a breeding tank look like a show tank. Even in a show tank, cichlids are going to breed. (Just try to stop them!) However, in a show tank we are more concerned with appearance. Hence, we may not like the looks of a number of inside filters in the tank. Thus, we may use a canister filter for filtration. These are quite efficient and provide filtration under pressure. Please be advised, however, that cichlids are active fishes of fairly large size, and they are going to provide lots of metabolic wastes. For that reason, we want to have a filter that can be changed at least every two weeks. So if you get a canister filter, get one in which it's easy to change or clean the media. Activated carbon is often a component of the filter media in a canister filter, so such devices do provide all three types of filtration, and they do so in an efficient manner.

Outside power filters fit on the side of the tanks. These are easily changed and can have enough different filter media to provide all three types of filtration. Those with a bio-wheel are particularly adept at providing biological filtration. A bio-wheel provides wet-dry filtration in a compact way. Wet-dry filtration is an attempt to provide the bacteria with the oxygen they need to do their good work. The bio-wheel provides that by a large rotating wheel that dips into the water and then rotates into the air, thus exposing a large surface to the air for gas exchange.

Trickle or wet-dry filtration created a sensation when it was introduced. It consists of a tower in which some sort of suitable habitat for bacteria is housed. Usually, this consists of "bio-balls," small plastic balls with ridges and crevices that provide lots of habitats for the bacteria. The water is pumped out of the tank through a prefilter (for the mechanical filtration) and then allowed to trickle down through the bio-balls for the dry part of the filtration. The water is collected in a sump, from which a pump propels the water back to the tank.

Although one of the ultimate biological filters, the wet-dry filter is expensive, and it has not been widely used in cichlid aquaria because cichlid people tend to emphasize regular partial water changes.

**Water Changes**

The fact is that it was cichlid people who pioneered the idea of regular partial water changes. The old idea was that there was something magical about old water. Cichlid and marine hobbyists helped break that taboo when the tanks in which regular partial water changes were made had better success when compared with those that had no regular partial water changes. This is true regardless of the type of filtration used.

The lesson here may be that filtration isn't that important in the cichlid tank, but I think it is a mistake to downplay filtration. Good filtration buys us time. Our water stays in good shape for a longer period of time. And most of us make trips, too, and we don't want to assign water changes to whoever looks after our fish for us. Besides, we never know when something is going to so command our attention so that we are less than exemplary about keeping to a water-changing schedule. One bit of apparatus that will make water changing easier, and thus more likely to be done, is an automatic siphon device that attaches to the faucet. Not only does it fill the tank with water, but it also removes it. These are easily attached and deployed, and they beat having to carry five-gallon

**Rockwork is desirable for many kinds of cichlids. This particular arrangement works well for African cichlids. For the larger Central and South American cichlids you would want an arrangement with more open spaces. Photo by MP. & C. Piednoir.**

**Driftwood offers a bonus for people who keep fishes that require reduced pH; it releases acids that help to slowly and gently acidify the water. Photo by MP. & C. Piednoir.**

Big fish need big tanks. A show tank like this filled with cichlids is an asset to any space. Photo by I. Francais

don't want to have the gravel be as fine as beach sand, for that is difficult to vacuum without getting some of the sand, too, since it is nearly as fine as some of the particles of debris.

## OTHER CONSIDERATIONS

Returning to the subject of tankmates for just a minute, it is worth mentioning that the same type of people who like cichlids also tend to like catfish, especially of the family Loricariidae (suckermouth catfishes). The bad part about this situation is that cichlids and catfishes both tend to like to lay out claims to the bottom of the aquarium. You don't need catfish for cleaning up in a cichlid aquarium, as most cichlids are quite adept at buckets of water! Naturally, you need to use special compounds for breaking down chlorine agents in the water.

## SUBSTRATE

Aside from ornaments, filtration, and tankmates, a special consideration for cichlids is the type of gravel employed in the tank. Most cichlids like to dig in the gravel, so there is little worry about anaerobic spots developing in the gravel without a subsand (undergravel) filter in the tank. My recommendation is to utilize a fine grade of gravel so that it is easier for the cichlids to dig, but only have a relatively shallow layer in the tank, of about two to three inches. The gravel bed will need to be vacuumed regularly, so you

Dark- and natural-colored gravels are best for cichlid tanks. The gravel size shown here is perfect for any cichlid aquarium. Photo by MP. & C. Piednoir

picking food from the bottom, and many species keep the gravel cleaned off pretty well, too. Hence, my recommendation is to leave catfishes out of a cichlid tank. I mean, I like catfish, too, but I try to keep them separated from cichlids as much as possible, and I match the species up very carefully.

In general, most tanks are going to be good cichlid tanks. Modern tanks come made with regular glass or acrylic, but the choice is one of personal preference. The lighting of a cichlid tank is also one of taste, but it is worth bearing in mind that some cichlids look good in bright light, but most are at their best in slightly subdued lighting. With that in mind, you may want to consider a light with a rheostat so that you can vary the intensity of the lighting. Since you won't be keeping plants, there is no reason to have the lights on when no one is around to enjoy viewing the tank. For that reason, it is best to keep the lights off when not needed, as having them on for protracted periods of time will encourage the growth of algae.

**A gravel washer is a very useful piece of aquarium maintenance equipment. Use it often! Photo by I. Francais.**

## FOOD

We are fortunate that these days there are many good prepared foods for fish species in general and cichlids in particular. An occasional feeding of live brine shrimp is great, but it is not a necessity, as nearly all species of cichlid prosper on the commercial foods, both dry and frozen. The only choice you need to make involves the size of the flakes or pellets that you feed, and that choice will be dictated by the size of your fish species.

**Brine shrimp nauplii at a few hours post hatch. This is very important first food for cichlid fry. Small adult cichlids are very fond of adult brine shrimp. Photo by B. Allen.**

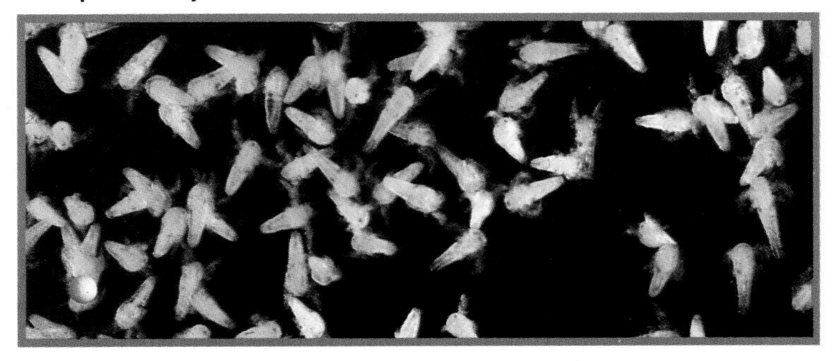

Usually it is not recommended that discus be kept with angelfish, but under some circumstances they can prove compatible. Basically, the discus should be larger than the angels, there should be plenty of plants, and the tank must be large enough to allow each its space. Photo by MP. & C. Piednoir.

# THE KINDS OF CICHLIDS

First, it may be of interest to us to be able to recognize cichlids. If we were going to specialize in cichlids, it would be an embarrassment not to be able to recognize a rare or unfamiliar species. Ichthyologists generally work with preserved specimens and utilize low power microscopes for examining the fish in question. But can we identify live cichlids and at least know that is what they are? Yes, we can, and here is what to look for.

**The extended fin rays on the dorsal fins of these *Apistogramma bitaeniata* contribute much to their glamour. While these fin rays are functional, no one said they couldn't be good looking as well! Photo by K. Knaack.**

## RECOGNIZING CICHLIDS

Remember that we said that cichlids had only two nostrils? That will be one of our main identifying points. If we get some marine species, that could confound us, as damselfish have similar features, and ichthyologists at one time considered them part of the same family. In any case, the two nostrils, together with a look at the fins and lateral line, should help us identify a fish as to the family Cichlidae. Cichlids have a broken lateral line in nearly all species. Many cichlids also have lateral line-like sensory pits on the head, but this won't help us determine that the fish is a cichlid, as many species have this feature. In cichlids, the pectoral fins and ventral fins are even. Again, this is a feature that many modern fishes have, but we are developing a composite picture here. Almost all cichlids have a continuous dorsal fin, with soft rays toward the posterior part of the fin. They all have hard spiny rays in the anterior part of the dorsal and the anal fin. This helps protect against predation, and it helps give the fins a strong structure to anchor the rest of the fin. The part with soft rays is often quite capable of undulations that help the fish to make delicate changes in position or to glide effortlessly through the water (in the "stealth mode"!).

So if we see a spiny-rayed fish with only two nostrils that has ventral and pectoral fins even with one another, we can be certain that it is a cichlid as long as it is not a pomacentrid (a damselfish of the family Pomacentridae)! If we know that the fish is a freshwater one, that fact alone eliminates our confusion. But some cichlids venture into the ocean in certain areas, too. If we were collecting in the ocean and came upon a new species in certain areas, we might have to dissect a specimen to be certain whether it was a damsel or a cichlid. Cichlids are also closely related to wrasses, parrotfish, and surfperches, but those species are easily told from cichlids because of the single nostril on each side of the snout that is peculiar to cichlids and damsels.

It is interesting to note that early scientists did confuse wrasses with cichlids, and that is how cichlids got their name. *Kichle* is a Greek word referring to thrush-like birds. They also used this name for several common wrasses in the Mediterranean because the fish they observed had certain behaviors of birds, swooping and feeding together. A species that looked birdlike (to the describer) was described as *Cichla* for the generic name in 1801, but cichlids were still not recognized as a separate family. In 1859, Bleeker recognized the cichlids as a separate family and named it Cichlidae after the genus *Cichla*. (Well, actually he called the family *Cychloidei*, but the endings for family names have changed since then.)

### WHERE ARE THE CICHLIDS?

Distinct types of cichlids occur in different places. In deciding what type of cichlid you want to keep, it may be useful to consider the types. The most profoundly evolved cichlids are in South America, and it is no accident that the discus and angelfish come from there and are excellent examples of cichlid species that have evolved into profoundly specialized forms. Cichlids don't dominate in South America, but they have certainly carved out niches for themselves. Conversely, they do dominate in Central America, not in biomass but in species. It is believed that cichlids were one of the first families to colonize the area as it arose from the sea.

More primitive cichlids are found at the tip of India and on the island of Madagascar. In fact, those cichlids may resemble the marine ancestor from which cichlids are believed to have descended.

The area in which there are the most cichlid species is Africa, and the main reason for that is that somehow cichlids have dominated the rift lakes of Africa, making them virtually giant cichlid fish bowls. Whenever you hear someone speak of African cichlids, they are nearly always talking about cichlids from one of the rift lakes. The

**Is this a pike cichlid? No, it's a wrasse, *Liopropoma eukrines,* a marine species that shares family ties with the cichlids.**

species from Lake Malawi are particularly colorful and active, so they are especially popular with those aquarists who want motion and color in a large tank for the living room, den, or office. However, the cichlids of Lake Tanganyika are particularly popular with advanced cichlid specialists and other hobbyists. This is because the cichlid species in that lake have much more variation. There are two reasons for this. One is that the lake is older and the cichlids have had more time to evolve to fill various niches. The other is that the cichlids are derived from more types of cichlid ancestors. For example, in Lake Malawi, all but one of the known species cares for the eggs and young via mouthbrooding. There are many different types of mouthbrooding in the lake, but there is none of the biparental care of the free-swimming fry for which cichlids are so well known. Conversely, in Lake Tanganyika, not only is there biparental guarding of fry, there is such extreme specialization among certain species that they are hardly recognized as cichlids. By contrast, the cichlids of Lake Victoria are mostly of the genus *Haplochromis*, and there is a certain lack of variation of body type, even though many species are quite beautiful.

There is one species of cichlid that occurs naturally in the United States, the well known Texas cichlid, but a number of species have established themselves in Florida, possibly as escapees from fish farms, and it is much feared that their impact has been severe on native

**This *Etroplus maculatus*, or orange chromide, hails from Sri Lanka. It is a gentle creature that favors slightly brackish water conditions. Some specimens are redder than others, but as always with cichlids, proper care will bring out the best colors available in a particular fish. Photo by M. Smith.**

**Paretroplus menarambo*. This fish is a native of Madagascar. The genus name, *Paretroplus*, means "alongside *Etroplus*," and they are indeed similar in many respects. Photo by M. Smith.**

species. Certainly many species from both Africa and Central America have flourished there, not to mention South American species deliberately introduced by Fish and Game personnel.

**PARENTAL CARE IN CICHLIDS**

Generally speaking, the cichlids from Central America and South America show the most complex and intense brood care, but this is not a hard-and-fast rule, as *Hemichromis* and *Tilapia* of Africa have some of the same intensity, while many cichlid species from Lake Tanganyika also demonstrate intense care. Of course, all cichlids provide some parental care; it is just that some species are more serious about it! In general, once again, the most colorful species are from Lake Malawi in Africa, but there are challengers not only from the other African lakes, but also from Central and South America.

**WATER CHEMISTRY FOR CICHLIDS**

Another consideration to take into account is the type of water that comes out of your tap. It is much easier to match your species of fish to your water than to match the water to your fish. If you have soft water, you may prefer to keep the cichlids of South America and West Africa, while if you have hard water with a high pH, the Central American cichlids and the cichlids of the Great Lakes of Africa are going to thrive in your tanks without your altering the water. For those who want to keep cichlids of the rift lakes of Africa but have soft water coming from the tap, rift lake salts are available from commercial enterprises at your aquarium shop. Such mixtures will help you reproduce the water of the lakes and help maintain the pH of your tank.

The cichlids of Central America are more adaptable, even though they prefer hard water and a higher pH. However, if you have extremely soft tap water that is slightly on the acid side, you may want to consider some of the rift salts even for them.

Before you decide upon which cichlid you want to keep, let's take a look at some of the typical cichlids from each area.

**_Apistogramma inconspicua._ This little female will take fine care of her fry while she can keep them all close to her. There are few sights more compelling to a fishkeeper than a cichlid, any cichlid, with a huge school of tiny fry.**

# THE CICHLIDS OF CENTRAL AND SOUTH AMERICA

Just so you will be familiar with the term, ichthyologists often refer to animals of the New World tropics (the Americas) as "neotropicals." So all the fish we discuss here are technically referred to as neotropical cichlids.

*Herichthys dovii*, or wolf cichlid. Don't mess with these two, whatever you do! Photo by D. Conkel.

It is interesting to note that as adaptable as cichlids are, they have not adapted to low temperatures. The original colonizing ancestor was apparently a tropical fish, and cichlids seem to lack the capacity to make that one adaptation— although there may be other unknown factors involved—as it is difficult to suggest in any way that cichlids lack any adaptive capability in light of their history!

It is with tongue firmly planted in cheek that I refer to Central American cichlids as "real cichlids." It is somewhat presumptive to refer to any group of cichlids as the "real McCoy." That is because cichlids have become so popular among tropical fish hobbyists that different groups of cichlidophiles are advocates for different groups of cichlids from different parts of the world. In fact, there is good-natured intramural warfare about which group is the prettiest or the most interesting to keep.

## SCIENTIFIC NAMES

Cichlid keepers become familiar with scientific names because popular names are often unreliable. For example, a species may be called a Texas cichlid in one part of the country and a "Rio Grande perch" in another. Nevertheless, even the scientific names can be a problem, and they are in a particular state of flux when it comes to many of the cichlids of Central and South America. Still, nearly all serious cichlid hobbyists learn the scientific names of the different species. With the popularity of the aquarium hobby, however, even ichthyologists will often suggest a popular name for a new species being described.

It is not difficult to get used to using scientific names. Generally speaking, we will only be concerned with species names, but let me quickly review how they fit into the scheme of classifying living things.

There are several levels of classification in biology. The most inclusive level is the kingdom. In fact, there used to be only two types of things in a kingdom: either plants or animals. New discoveries in biology have led to at least five kingdoms now, and some biologists recommend six, with one for viruses. (Since viruses don't respire, some biologists refuse to consider them living things, and yet they are grouped and classified in biology.) The five kingdoms presently accepted by all biologists are Monera, Protista, Fungi, Plantae, and Animalia. These terms refer to bacteria, the larger single-celled organisms with the genetic material in a nucleus, fungi (including mushrooms), plants, and animals.

The levels of classification are the following: kingdom, phylum, class, order, family, genus, and species.

We can see that each group is less and less inclusive; that is, it is more restrictive of species. Thus, if we take the human species (that is so often incorrectly referred to as a "race" in the popular media), we are in the animal kingdom, the phylum of chordates (Chordata), the class of mammals (Mammalia), the order of primates (which includes monkeys, lemurs, and apes), the family Hominidae (of which we are the only surviving members, although there are several fossil species), the genus *Homo*, and the species *sapiens.*

This system of classification has served very well ever since its inception. Of course, scientists tinker with it a bit to meet their needs. They will thus utilize groupings in between the classifications. For example, there are subphyla and superfamilies. Such classifications are merely a convenience, a way of dealing with animals so that they fit in the system, but it is more realistic to consider some very similar animals (or whatever) in "superfamilies," for example. These would be animals that are too dissimilar to place in the same family, and yet they have some common traits that tie the families together. In the case of cichlids, they are grouped with the marine families Pomacentridae, Labridae, Scaridae, and Embiotocidae, to form the superfamily Labroidei. It is just a matter of stretching and bending the classification system to accommodate the fact that there are often not clear distinctions between categories. After all, the whole classification scheme is simply an artificial construct of humankind to enable us to better understand the universe. The same could be said of certain features of physics and mathematics.

As I said, we are primarily interested in species names. They consist of two names, the genus name (or generic

***Heroina isonycterina*, a new species of cichlid from Ecuador and Colombia described by Dr. Sven O. Kullander in 1996. Photo by M. Smith.**

***Hypsophrys nicaraguensis*, a splendid cichlid that was misplaced in the genus *Copora* for a while but is now returned to the name given by the great ichthyologist Louis Agassiz some 150 years ago. Photo by MP. & C. Piednoir.**

name) and the species name. Thus, we are *Homo sapiens*. That means "wise man." (In Latin, the adjective follows the noun.) Of course, it is nice to be able to do the naming, as we can give ourselves a good name! Notice that the name is in italics and that the generic name is capitalized, but the species name is not. A common error in the popular press is to fail to italicize the names and to capitalize both names or to capitalize neither.

With most Central American cichlids, the generic name has always been *Cichlasoma*, but therein lies a story of some controversy. If the reader will remain steadfast for just a moment here, we will review that quickly and then get to the fish.

*Cichlasoma* is a scientific name that has been used for many years to cover cichlids from South America and Central America. It has been known all along by ichthyologists who specialize in cichlids that a diverse group of animals has been covered by that designation—too diverse to be included in one genus. One temporary solution was done by the great ichthyologist Regan, who broke the Central American groups up into what he called "sections." Other ichthyologists have generally treated those sections as subgeneric groupings. Although ichthyologists who took a special interest in cichlids acknowledged that work needed to be done on the *Cichlasoma* genus, none was inclined, apparently, to tackle the great work involved in splitting up this group of over eighty species.

In 1983, Dr. Sven Kullander began the work that so many had avoided for so long. He restricted the genus *Cichlasoma* to just twelve species in South America. This, in effect, "orphaned" all the other *Cichlasoma* species. The question was what to call the other species until Kullander, or someone else, did the work to break these various cichlids up into different genera. One solution was to use the next available name that had been used for *Cichlasoma*, which was *Heros*. The problem with that was that the name *Heros* was soon utilized for a distinct group of cichlids, so that, once again, the other cichlids of the *Cichlasoma* group were orphaned. The next available name is *Herichthys*. The "available" names are generic names that were once used for cichlids of the *Cichlasoma* group, but they were used

**A male *Herichthys festae*, the red terror. This fish is well named! A real beauty to be sure, but one of those meanies that bullies its neighbors and beats the wife. Photo by MP. & C. Piednoir**

after *Cichlasoma* had been used. For that reason, *Cichlasoma* had priority. However, once *Cichlasoma* was restricted to just the species in South America, the junior synonyms, such as *Heros* and *Herichthys*, could be used. *Heros* had priority, but now it has been restricted to the *severum*-like fishes of South America, so *Herichthys* is now available. The trouble is that *Herichthys* is very likely to be restricted to the group of cichlids that the Texas cichlid belongs to. That means that we will once again be looking for another name. For that reason, I propose to retain *Cichlasoma* for most species, but I will use the "sections" of Regan, as they are presently applied, in parentheses, as these very likely will eventually become the generic names for the species so indicated.

**_Herichthys carpinte_, the pearlscale cichlid, is a close relative of the Texas cichlid (if not actually the same fish from a different location). These big busters are as tough as they look and should only be kept with fish that can hold their own. Photo by MP. & C. Piednoir.**

The fact is that we are not in bad company here, as many ichthyologists still use *Cichlasoma* in the scientific literature, waiting for all the revisions to be made and for general acceptance of them before switching names. This is conservative, and it is the easy way out. And those are good enough reasons for me!

I hope that all this turmoil with scientific names has not soured the reader on using them. With practice, these names become old friends. And I suspect that is one reason even scientists are reluctant to drop *Cichlasoma*, as it has been around for a long time and has a pleasant connotation for many of us.

In any case, scientific names have much value. They are used universally all over the world. The pronunciations may vary, but the names are written the same everywhere on the sphere. Also, scientific names show relationships. Even the old *Cichlasoma* grouping showed that the fishes had a relatively close relationship, even if it was not close enough for most ichthyologists to accept for many of the fishes in the genus.

The overwhelming reason for using scientific names, though, is that cichlidophiles use them. If you want to be considered a genuine full-blown cichlid person, get used to them!

## CENTRAL AMERICAN CICHLIDS

Before listing some species, let me comment on species that are particularly good for newcomers. One of the things that you want to be able to do is to have a fish species that unfailingly spawns and tends and protects its young. Raising cichlids is not as easy as it can sometimes seem. Sometimes things go horribly wrong and the fish eat their own young and then one parent kills the other. Such fish are either behaviorally defective or they are reacting to the artificial confines of the aquarium. Also, some species take a long time before they spawn (often because they take at least two years to mature) or are infrequent and temperamental spawners. Such fish species are cakes and ale for the experienced cichlidophile, but they can discourage a newcomer.

There are two species that I recommend for beginning with cichlids of Central America. One is the convict cichlid,

***Cichlasoma (Archocentrus) nigrofasciatus*, the convict cichlid, is a fish that will astound newcomers to the cichlid hobby. It is precocious, spawning at an early age, and exhibits all manner of clever cichlid behavioral quirks. Photo by D. Conkel.**

**The marbled convict cichlid, a new morph of the old favorite. Photo by Dr. W. Leibel.**

***Aequidens coeruleopunctatus*** **is a cichlid that will quickly eat the young if they are threatened. While this may sound barbaric, it makes good sense for the parents. Spawning takes a lot of energy and this cannibalism keeps the fuel needed for the production of the next spawn in the family where it will do the most good. Photo by D. Conkel.**

**The red devil is an aggressive, capable fish that should be kept by itself in a large aquarium. In its native waters it is often caught on hook and line by anglers. Photo by D. Conkel.**

*Cichlasoma (Archocentrus) nigrofasciatus*. Some would disagree with me on this species, as it is quite aggressive; however, it is my view that the aggression is simply secondary to the excellent care that it gives its young. In any case, it is an unfailingly good parent, and it spawns early and at a small size. The other species is the common firemouth cichlid. (I am taking into account the availability of species, as I could recommend some hard-to-find species that might be even better, but...they are hard to find!) This species may take a year to reach spawning maturity, but it is not as aggressive as the convict cichlid and is a good parent. Another favorable aspect is that both parents give equal care to the young and cooperate admirably.

I'll include them in the list of species. Just give them special consideration!

- *Aequidens coeruleopunctatus*

Although this species has no popular name and it is not particularly colorful, it is worth keeping as the only representative of its genus in Central America. It is found from Costa Rica to Panama. Although a good parent, it is relatively peaceful otherwise, much like the *Thorichthys* group. The males don't quite reach six inches in length.

- *Cichlasoma (Amphilophus) labiatus*

Red Devil

When these cichlids first arrived in the country, dealers nearly put them in ocean water, as they resembled the Garibaldi, a large California damsel, in color. Exporters had actually mixed up red devils with red Midas cichlids,

*Cichlasoma (Amphilophus) citrinellum*. For that reason, the Midas cichlid is still called a red devil in some circles, but it tends to be golden, rather than red, and it has a less pointed snout. The large lips of the red devil tend to regress under captive conditions for some reason that is not completely understood. These cichlids were not named red devils without reason, as they are quite aggressive and have the dentition to back it up. Nothing could seemingly be kept with them, but it was later discovered that they could be kept (in very large tanks, of course) with a number of other cichlids while growing up—any of the tough ones!

**The Midas cichlid is well known for its dramatic nuchal hump. Photo by D. Conkel.**

With all of these cichlids, the best way to breed them is to get about six juveniles and allow them to pair off naturally. Of course, once a pair spawns, you will have to remove the others—unless you have a 2000-gallon tank! The juveniles are gray in coloration and then begin to turn speckled before they turn red. Interestingly enough, the red coloration is quite variable, and it is often fringed with black (the most beautiful pattern, in my opinion), and a fascinating fact is that in the wild many of the individual specimens stay gray. This fact has been much studied by scientists.

The male can reach nearly a foot in length, but it takes about three years to attain such a size. The pairs will spawn when only six inches long. Obviously, this is not a cichlid for beginners, but it is so spectacular and is so often kept, even by the uninitiated, that I could not avoid listing it.

**Young males and females do not sport the same degree of headgrowth as the older male Midas. Photo by M. Smith.**

***Archocentrus sajica*** **is a handsome fish that rarely measures over 3 inches. For a cichlid, this fish is a pussycat—until there are babies to defend! Photo by J. Elias.**

- *Cichlasoma (Archocentrus) nigrofasciatus*
  Convict Cichlid

This is a cichlid that has often been called the "missionary cichlid," as it so often entices a neophyte into the cichlid hobby. It is not that the fish is so beautiful, but it is unerringly a good parent, and on the very first try, too! Further, it spawns at a young age, and it seems to be either in the process of spawning or raising young during most of its life span. A further favorable attribute of these cichlids is that they spawn at a very small size, with the male barely over an inch long, and the female barely reaching an inch in size.

Of course, these are quite common cichlids, but they never lose their appeal. A recent innovation among cichlidophiles is to try to collect the various color morphs. Since this species is spread over a wide area of Central America, there is considerable variation in coloration dependent upon the locality where the species is found. There are particularly red color forms that are found in Honduras. In these specimens, the red is found on both male and female. But on the female, it covers more of the body, producing a flaming belly region and red all the way up to the dorsal fin. In the more common types, the male shows no red, but the female sports a copper coloration in the ventral (belly) regions.

All of the *Archocentrus* groups are known as fierce protectors of the young, and this species displays it as much as any. For example, a famous story involved a surplus of convicts being fed to a piranha in a Los Angeles fish shop. Some of the babies escaped and matured to spawning size. The resulting pair nearly chased the piranha out of its own tank, even though it looked like a whale as compared to the cichlids, and the shop owner had to move it for its own safety.

The *Archocentrus* group is one that feeds on the bottom, picking up foods opportunistically. The food can vary from invertebrates inhabiting the substrate to a mixture of organic matter that ichthyologists euphemistically call "grut." The parents feed the young off a specialized body slime that is secreted by both parents. This supplying of supplementary "milk" is very common among Central American cichlids. The parents also supply food for the young by chewing up food that is too large for them and spitting it out in the school of young. They also "sweep" the bottom with their anal fins to stir up any possible food for the young. All these behaviors, though typical of many cichlids, are what have endeared this particular species to aquarists around the world.

- *Cichlasoma (Archocentrus) sajica*

There is no popular name for this cichlid, as it was recently described, and it is very likely the smallest of our Central American cichlids. The species name *sajica* is the Spanish rendition of an Indian name, so it is pronounced "saw-HEE-ka."

Although not particularly colorful, it has interesting markings. And it has nearly all the behaviors of the convict cichlid, and it changes colors when spawning. (Convicts merely darken slightly.)

- *Cichlasoma (Archocentrus) septemfasciatum*

  Copper Cichlid

This is another cichlid that changes dramatically when spawning. The popular name comes from the copper-like coloring that often covers the body of the female. Something that all of the *Archocentrus* group share is that the female is pretty much in charge of the eggs before they hatch, while the male guards the perimeter. Once the eggs hatch, the male aids in moving the young to a hole that is dug. As a matter of fact, several holes are dug in the process of the young becoming free swimming. The period of time involved is about four days, and the reason for moving the young from one hole to another has been debated. One early theory was that it made a moving target of the young, as they didn't stay at a permanent address. Although that may be part of the reason for the behavior, an overwhelming pressure is to keep the young clean. As they are transferred via mouth by the parents from hole to hole, they get cleansed of debris and any bacteria or fungus that may be attacking them or growing as a culture in the debris.

Through a sorting error, this species was originally introduced into the United Stated as *Cichlasoma spilurus*, and it is found in the older aquarium literature under that name. When the species spawns, the female develops a dark mask that looks very much like a knight with his hood pulled down, ready for battle. While not a knight, a guarding copper female is truly ready for battle!

- *Cichlasoma (Archocentrus) spilurus*

Although superficially similar in color to the convict cichlid, the stripes are of a different pattern, and the males often have a yellow coloration to the body. An interesting point among the very similar *Archocentrus* listed here is that the farther away they get from each other's range, the more they resemble one another, with the individuals living sympatrically with other species showing the most difference in coloration. Biologically this makes sense, as hybridization from a failure to recognize different species would spell the end of the line for the genetic material of the species involved, as hybrids seldom prosper or reproduce successfully in the wild.

- *Cichlasoma (Herichthys) octofasciatus*

  Jack Dempsey

You can tell by the popular name of this fish how long it has been in the hobby, as Jack Dempsey became the heavyweight boxing champion more than 80 years ago. Most people these days simply refer to the fish as "Dempseys." And the fact is that the fish is not nearly as aggressive as the red devil, for example, and some of the others. However, it is durable and tough enough to be kept with such fish in a tank that is reasonably large. The species is found on the Atlantic slope, from Mexico to Belize.

Juveniles look primarily black, without the blue

***Archocentrus septemfasciatum* is available in many color morphs, all of which breed freely. This results in a good aquarium fish that presents a spectrum of colors and patterns. Photo by D. Conkel.**

spangling that really "makes" the appearance of this fish. It takes patience to see this fish at its best, as it takes slightly over a year to fully mature. And yet I have known of small specimens, barely three inches long, to breed in captivity. How rare this is would be difficult to say, but it most assuredly is non-existent in the wild.

Like all other Central American cichlids, this one prefers slightly hard and alkaline water, but again like the others, it is quite adaptable. It is a long-time favorite and deservedly so.

Incidentally, this species is only placed in the subgenus *Amphilophus* with a shoehorn. It is something of a stretch to put it there, so I have left it without a subgenus. Ichthyologists may eventually place it in a genus of its own, as it doesn't fit clearly into any of the proposed genera or subgenera. The males reach a length of about eight inches.

- *Cichlasoma (Thorichthys) aureus*

Golden Cichlid

Actually, there are several popular names for color variations of this quite charming cichlid, including "gold flash" for one of the variations. The most popular in the aquarium trade of this group is the firemouth cichlid, and all of the *Thorichthys* group greatly resemble one another. They are substrate-sifting, but they don't run the gravel out through the gills, as is so typical of *Geophagus*; rather, they turn the sand over in the mouth and then spit it out. In the wild, they feed primarily off invertebrates that they find in the sand, mostly insect larvae. In the aquarium, they take food opportunistically, and they are good feeders and hardy aquarium fish.

*Thorichthys* has been elevated to full generic status by some ichthyologists, but I am staying conservative here. These cichlids could best be described as average in size, as in several species the male reaches a length of six inches, with the female about five inches. But these cichlids are quite mild, as compared to cichlids in general. True enough, they are fierce protectors of their young, but they are not as capable of inflicting as much damage on other fish as many other cichlids, and when not spawning, they are more quiescent than most cichlid species. I have kept them with angels, for example, something that could not be done with the majority of cichlid species.

There are two main color variations of this species, with one being known as the "blue flash," sporting primarily blue coloration, and the other the "gold flash," showing more reds and golds. The fact is that some red (or gold) and blue is shown in each variation, but the division is understandable once you become familiar with these fish.

There has been a renewed interest in this species because of recent new importations and because of the fact that it is relatively new to the hobby. The *Thorichthys* genus is popular enough that some cichlid specialists tend to specialize in just this group.

- *Cichlasoma (Thorichthys) ellioti*

Although there is no popular name for this species, it

**A spilurus and a firemouth face off in a territorial dispute. Photo by J. Vierke.**

has become particularly popular with cichlid specialists in recent years for the same reasons given above with *aureus*. This species is found in Eastern Mexico from the Rio Papaloanpan to the Rio Coatzacoalcos systems.

**The Jack Dempsey is well known for its ability to defend itself but most people don't realize what a stunning fish it is in its adult manifestation. Photo by MP. & C. Piednoir.**

One of the interesting aspects of the *Thorichthys* group is that the male and female share duties in the care of the young. That is, the male takes his turn at fanning the eggs, while the female guards the perimeter or feeds briefly. Another trait of interest is the spot on the operculum (the gill cover). This apparently is an eye spot that is used in frontal displays. That means that the spots are enlarged when the fish confronts another fish of the same or different species. The gill covers and a membrane below the neck flare out, much like the ruff on a gamecock, and make the fish look larger. This is a frontal threatening display, and it has the advantage of sporting two extra eyes, and both of them are large. This jibes well with information ethologists have gathered that animals seem to be innately afraid of eyes, and the bigger they are, the more frightening they are. (Just think of us and of horror movies, with large eyes glowing in the dark.)

In any case, with this group of fishes, the displays help drive off other fish species without actual aggression. That may be one reason that this group is of a somewhat milder temperament.

- *Cichlasoma (Thorichthys) meeki*

Firemouth Cichlid

This species has long been a favorite in the aquarium, even though it is not suitable for the typical community aquarium. The reasons are obvious. It has a distinctive shape, and its coloration is quite pleasing also. It has always been one of my favorites, and the fact that it is common has never bothered me. Apparently, it hasn't bothered other hobbyists either, as it has remained consistently popular. The fact is that it can be kept in a community aquarium until spawning time arises. Then the other fish are driven off mercilessly. The fish spawn when the male is a mere three inches, with the female being about two and a half inches in length. However, the male can eventually reach nearly six inches in length. As compared to many other cichlids, the pairs seem to stay together well in the home aquarium, with no spats in between spawning.

This species's bluff may be even more impressive than that of the other ones in this group. That is, the frontal threatening display may be more effective. That is because of the red coloration that is more intense in this species, especially toward the front. For some reason, red coloration is also intimidating to fish, a fact discovered by ethologists.

Another interesting trait is how easily the pair recognizes one another. I once had the living room tank full of firemouths, and a pair had spawned. It was a large tank, so I left the other fish in. As far as I was concerned, all the firemouths looked alike. However, it was particularly amusing when the female was on guard and the male returned to the territory, approaching her from the rear. She whirled, ready to attack, but was immediately appeased upon recognizing the male.

There are color differences in this species, too, depending upon the locality. The species ranges from southern Mexico to Guatemala, being found primarily on the Yucatan Peninsula. Some species have more blue spangles, and with others there is more emphasis on the red. And it should be mentioned that there is some variation in the coloration in a single population of this species in one locality, with some specimens showing a brighter red than others. The most beautiful, and therefore the most popular, specimens are those that show a balance of bright red with lots of blue spangles.

- *Herotilapia multispinosa* Rainbow Cichlid

This fish is a candidate for the smallest of the Central American cichlids, with the males only reaching three inches, or only a little more. This is another one of those species that seems capable of defending its young without committing total mayhem on its tankmates. This fish is distinct from other Central American cichlids in that it has tricuspid (three-pointed) teeth. These enable it to more easily harvest filamentous algae, which make up a good portion of its diet. In the aquarium, it will take all foods, but some vegetable matter should be a part of its diet, even if it is only dry foods designed for herbivores.

- *Neetroplus nematopus* Poor Man's *Tropheus*

This species is somewhat analogous to the *Tropheus* of Lake Tanganyika, in that it lives off the algae that grow on the rocks in the lakes and rivers in which it is found. It somewhat resembles *Tropheus* in shape, too, and this fact has resulted in the species being dubbed with the popular name "poor man's *Tropheus*." This makes devotees of the species, or of Central American cichlids in general, bristle, for they feel that this cichlid is more than the equal of any *Tropheus*, most especially in behavior.

To be fair, many *Tropheus* are more colorful, and they will probably always be high priced because of the fact of their small spawns. In addition, *Tropheus* can be difficult to spawn in captivity. In the case of *Neetroplus* it is just the opposite. Just try to stop these guys from spawning! They run the convict cichlid a close second on propensity to spawn and the efficiency at protecting their young. This fish is found from Costa Rica to Nicaragua in lakes and rivers, including the Great Lakes of Nicaragua.

An interesting aspect to this fish is its reversal of colors at spawning time. The dark stripe down the side becomes quite white, and the body becomes nearly jet black. So in spawning colors it really does resemble *Tropheus duboisi* adults.

**Comments**

If you decide to start with my recommendation and keep some firemouths and convicts, please be aware that these species can be kept together even though the convicts are more aggressive. One reason is that they are quite different in appearance and another is that the firemouths are a little larger than the convicts, especially at spawning time.

**_Herichthys spilurus._ When a pair fights over the eggs, it may be advisable to remove the aggressor, male or female, depending on species. Sometimes the male is the better parent and should be the one left to raise the brood.**

## SOUTH AMERICAN CICHLIDS

- *"Aequidens" rivulatus*

Green Terror

This is one of the species that was excluded from the new definition of *Aequidens*, so quotation marks are placed around the generic name. To add to the confusion, the true "*Aequidens*" *rivulatus* may not be the one that is extant in the tropical fish hobby. The other one is similar in appearance and disposition, but the other has more iridescent green scales and it hails from the Pacific slope while the one in the hobby was collected in eastern Ecuador. The species reaches about ten inches in length, but it will spawn at a much smaller size, from five to six inches. The best way to spawn the fish is at that size, as you can obtain six individuals and let them pair up naturally. Otherwise, you will have to use a grid divider for a pair, as the large ones are hard to pair up.

***Neetroplus nematopus.*** **This fish is sexually quiescent. When excited, it really does color up to look like a *Tropheus*.**

**Right: *Aequidens rivulatus*, the green terror, is a gorgeous fish best kept by itself or only with good friends, as the common name "terror" is never applied lightly in the fishkeeping hobby. Photo by MP. & C. Piednoir.**

**Below: *Thorichthys aureum* is often called the blue flash and with good reason, as this fish is speedy and spangled with iridescent scales. Photo by D. Conkel.**

*Apistogramma agassizi*, the spade-tail apisto, in spawning coloration is an awesome little fish.

An *Apistogramma bitaeniata* female with fry is not to be messed with!

Once the fish attain their ultimate size, you very likely may need the grating anyway to protect the heater and other important fixtures in the tank. A single member of the species may be kept as a pet all by itself in a tank, as these animals are quite intelligent and learn to recognize their owners and even perform tricks for food. You will need great patience for the latter activity, but I have seen it done. Large specimens are also at their best in appearance and coloration.

Interestingly, the original fish which were brought in and earned the name green terrors had vertical fins with ivory margins. If your fish have orange margins on the caudal, dorsal, and anal fins, they will be relatively peaceful as compared to the original green terrors, but they can still be a handful!

Obviously, these fish are extra trouble to keep, but the species is quite fascinating in behavior, and its beauty can be jaw-dropping.

- *Apistogramma agassizi*
  Spade Tail Apisto

This fish was described scientifically in 1875, and it occurs throughout the Amazon, right up to the base of the Andes Mountains. This is a species that has been a staple in the tropical fish hobby among those who like small cichlids. All the known *Apistogramma* species spawn in a very similar way. The female turns a yellow color and guards the eggs by herself and takes primary care of the young, with the male patrolling the perimeter. Even during a spawning, the fish are unlikely to harm other inhabitants, as these are such mild fish that even a female guppy may challenge the fry-guarding female. In nature, most of the *Apistogramma* species are harem spawners, with one male presiding over the territory of several females, so it may be best to keep one male with several females in a community tank.

All apistos are found in especially soft water in the wild, but I have successfully spawned many species in San Diego water, which is best described as "liquid rock," so obviously they are versatile. Still, I may have been quite fortunate, and the reader is advised to provide soft water for spawning at least. But there is no question that *Apistogramma* species can prosper in hard water if you aren't worried about breeding them.

In most of the apisto species, the male reaches about three inches in length, with the female reaching about an inch to two inches. They spawn at a much smaller size. The apistos are short lived for cichlids, only living for about three to four years. There are a great many species of them, but most of them are relatively colorless and only appreciated by the real apisto devotee. In any case, I'll list just one more here.

In spite of the fact that these fish are small, the young are able to take newly hatched brine shrimp once they are free swimming.

- *Apistogramma bitaeniata*

Although this is a popular fish, there is no popular name for it, but it has gone by many synonyms. There was a time that all the apisto species that had the extended dorsal rays were known as *Apistogramma* U1 or U2 or U3, including

***Herichthys festae*, the red terror, is a big mean cichlid that brightens considerably when in spawning mode. It also gets even meaner than usual. It is best kept with a mate or some other big, tough cichlid. Photo by Dr. H. Grier.**

***Cleithracara maronii*, the keyhole cichlid, is shy and easily bullied by other cichlids. It is easy to spawn and an excellent aquarium fish. Photo by MP. & C. Piednoir.**

Pike cichlids are renowned for their capacious mouths. Illustrated here is an unidentified *Crenicichla* sp. in the process of swallowing an unfortunate *Aequidens.* Photo by Dr. H. R. Axelrod.

*Crenicichla notophthalmus.* This fish will grow to six inches. The eye-spot pike is slightly less predatory than the other pike cichlids, but it is still a long way from being a community-tank fish. Photo by H. Mayland.

*Herichthys ellioti* is from eastern Mexico. It sifts the sand for invertebrates and so benefits from the inclusion of live adult brine shrimp in the diet. Photo by D. Conkel.

even one U4. The "U" stood for "unknown," as the genus was recognized but the species was not known. This was back before the U2 spy planes became famous! Then Meinken described this species as *Apistogramma kleei*, and the species was known for a while in the hobby under that name. As it turned out, Meinken had not been able to review all the preserved material, as the species had been described as *Apistogramma bitaeniata* in 1936, and that name took precedence. (The earliest name used in a valid scientific description is the one that has priority according to biological convention.)

- *"Cichlasoma" festae*
  Red Terror

Here is another fish that takes a fanatic to keep it, but it brings enough attractive qualities to inspire such fanaticism. For one thing, a specimen in good condition can rival nearly any fish in majestic appearance and spectacular coloration. Second, the species has behavior that is not only fascinating but is downright spectacular. This is the type of fish that when it is defending its young will leap out of the water at a hand that is placed over the surface of where the fry are located! Still, a single specimen can be kept in a community tank of large and rough cichlids, but the tank should be quite large, of at least two hundred gallons capacity. That doesn't mean that the fish can't be spawned, but it is best done when they are relatively small. Pairing at full size is quite tricky, for the fish reach a considerable size, easily ten inches in length, and they are quite formidable. I once placed an adult male and several females in a 250-gallon tank, thinking that a relatively safe situation. Since there were so many females for the male to choose from, I reasoned, he wouldn't single out one female for bullying. Well, I miscalculated that time. A pair formed almost immediately, and they turned their defenses on the other females, trying to drive them away. And they were dead on my return late that day.

As usual, the best way to get a pair for spawning is to invest in six young and let them grow up and pair up naturally. They will spawn at a size of less than six inches, but a full-grown compatible pair with young is quite a sight to behold. The huge, beautiful pair acts as though it will eat the dog if it comes close to the tank. And even the aquarist is tolerated within narrow parameters.

Although the fish is well worth keeping, it comes honestly by its name "red terror." That is, it truly is really red, nearly glowing with its own light when in full spawning color, and it is truly a terror, too, as a parental pair will countenance no fish in the same tank with them, regardless of the size of the tank. It and the green terror were first imported at about the same time from the same general locality, the eastern part of Ecuador.

***Microgeophagus ramirezi*, the ram, is one of the most popular and well known of the South American dwarf cichlids. It does best in the soft, acidic water of its natural habitat. Photo by MP. & C. Piednoir.**

- *Cleithracara maronii*
  Keyhole Cichlid

This fish was first described as *Acara maronii* in 1882, and it is found in the Guianas. It reaches a length of about four inches, but it is one of the most gentle of all cichlids. The parents share the care of the young, but they are quite

***Geophagus hondae* is usually a relatively peaceful fish. The mouths of the fishes of the genus *Geophagus* are quite exaggerated and extend to unusual positions.**

*Nannacara anomala.* This small beauty is often kept in the peaceful community tank, where its subdued good looks are much appreciated.

*Apistogramma agassizi.* This little fish does very well in a small planted aquarium. Condition with small live foods for breeding. Photo by H.-J. Richter.

gentle in their protection. In fact, they should not be kept with other cichlids other than those that are small and gentle. The fish will be found in the older literature as *Aequidens maronii.*

- *Crenicichla multispinosa*
  Pike Cichlid

Although the *Crenicichla* species are deadly ambush predators, many species make good aquarium residents for the special tank, the main requirement being that all members of the tank be too large too swallow. Although the fish are excellent parents, they aren't overly aggressive as most cichlids go. Once sexual maturity is reached, the males can be told from the females by their larger size (up to eight inches) and more spangles on the body. But the females have the best coloration, sporting a bright red in the belly region.

Nearly all of the *Crenicichla* species are cave spawners, and aquarists use flowerpots and PVC piping to accommodate them. Although they are efficient fish predators, they will eat dry and frozen foods.

- *Geophagus jurupari*
  Jurupari

The name jurupari is a native name, so it is not a patronym and is not pronounced with the long "i" at the end. This is one of those species that has undergone revision and has been placed in the genus *Satanoperca.* Because of the fact that the species has been so well known by the old genus and the ichthyological community may not accept that name change in the long run, I have used the old name. It is only fair to say that many ichthyologists who know the species quite well accept the name change. (It is my hope that those new to scientific names won't be put off by all of the name changes. The changes are a healthy sign in that they reflect that a lot of research and study is being done with these animals. Thus, the changes in names reflect a refinement of our scientific understanding of these fishes.)

This oscar (*Astronotus ocellatus*) shows all the earmarks of being stunted. When a fish is kept in an aquarium that is too small to permit proper growth, the gills curl and the body is shortened. Be sure to keep your fish in a tank that is large enough for them at maximum size.

These fish are amazingly adaptable to water quality and are found throughout the Amazon, but it has to deal with different water parameters in some coastal areas. This species can reach nearly a foot in length, but it takes it at least three years to do so. It is not aggressive with other cichlids, so its size helps protect it from some of the smaller more aggressive cichlids.

- *Microgeophagus ramirezi*

  Ram

This fish was named after its collector, Manuel Vicente Ramirez. Since Americans butchered the scientific name, the popular name of "ram" somehow came about. There is also a golden variety, but, as so often is the case, the wild color is the most beautiful. Although the colors are somewhat subdued, the fish is truly beautiful when in full coloration, and its appearance has a real touch of class.

In spawning, both parents clean off a rock and both of them care for the young, just like some of the large cichlids. However, they are only about two inches long, at the maxi-

This is a 4-year-old wild-caught oscar. Many people keep oscars singly in a show tank where they impress with their intelligence and playful behavior. Photo by MP. & C. Piednoir.

This is a long-finned albino oscar. Oscars are South American cichlids that are wonderfully hardy and personable. They should not be kept with fishes small enough for them to eat, but are not extremely nasty despite their considerable bulk.

The angelfish is easy to keep and breed and is the cichlid most frequently kept, and kept well, in the community tank. While it is still a cichlid—and exhibits a full range of cichlid behavior—it is considered peaceful enough to be kept with small fishes. This is sometimes a mistake.

mum, and they are quite gentle, so they are no threat to the other members of a community tank. And they will be one of the favorite additions to the tank. Although I spawned the species in our hard San Diego water, I had the best success in water that had been softened.

- *Nannacara anomala*

This species is very similar to *Apistogramma* species in behavior, but it is considerably different anatomically. I called them "animated watermelons" when I first saw them. They look merely transparent brown under bright light. But when they are in good condition and the lighting is not too bright, the male has a bluish iridescence to him. These fish spawn in an identical manner to apistos, except that the female does not turn yellow. Once again, this is a species that can spawn in a community tank without terrorizing and killing the other inhabitants, although they will be chased away from the spawning area.

- *Pterophyllum scalare* Angelfish

This exotic fish is thought of as common now, but it certainly created a sensation when it was first imported early in the century. And you can bet that they were expensive then, as they had to be shipped in large containers by freighter all the way up from the Amazon. They were not easy to induce to spawn either, but their descendants, who have been captive-bred for many generations, are not so difficult. In contradistinction to most other cichlids, angelfish like to stay up in the water column, although they will pick up food from the bottom, looking unbelievably stately as they do so.

Angelfish spawn on plants or tree limbs in the wild. When the young hatch, they move them to other plants (instead of putting them in pits, as is the case with so many other cichlids), and they transfer them from plant to plant. When the young are free swimming, they will take newly hatched brine shrimp, and both parents care for them. Unfortunately, angelfish are raised commercially by removing the eggs from the parents, so many angels may not be effective parents, since the bad ones have not been selected out in a Darwinian manner, as they would have in nature.

Although there are many fancy varieties that have been bred in the aquarium, including gold varieties, the natural form still looks the best to those of us who want our fish as they are found in nature. There are two other species in the genus, *P. altum* and *P. leopoldi*, but they are difficult to tell apart unless you have a practiced eye. The males and females in all three species are also quite similar in appearance, requiring an expert to tell them apart, and even our expert may be wrong half the time!

- *Symphysodon discus*
  Discus

Along with the angelfish, this species is generally considered the most exotic of tropical fish species, but many cichlidophiles would disagree. There are believed to be at least two species of discus, with several subspecies. The other species is *Symphysodon aequifasciatus*, but there are ichthyologists who will argue that there is only one species. In that case, it would be the above, as it was the first one described.

**Discus are now being bred in a wide variety of color forms. Keep their water clean and provide high quality foods and your experience with discus will be beyond your wildest expectations. Photo by MP. & C. Piednoir.**

There are two types of discus hobbyist. The most common are the ones who breed fancy varieties. I am not in that camp myself, as I find most of the fancy varieties inane. For example, one fancy variety is an elongated type. Isn't that absurd? The charm of the fish was always its disk-like shape and majestic manner, and some crazy discus hobbyists decided the shape should be different. Oh well, if you like that sort of thing, I won't complain, but the other camp of discus hobbyists is the one I like the most. They want their discus as close to the wild types as possible, and they even make exhibitions down to the Amazon to capture wild specimens for diversity in the captive genetic pool.

**Comments**

There are so many fabulous and fascinating cichlids in the New World that it is a frustration not to list more of them. However, you can get a sample of what is available from this, from the gentle to the untamed. They all have their charms, and they all have their adherents. If your conditions are right, your cichlids will invariably spawn. That means that the other fish in the tank are in trouble. That is why you need lots of rockwork for the truly aggressive species. You may need to remove the other fish to another tank or place a partition in the tank to protect them from the wrathful parents. (Such necessity is the dark side of cichlid keeping!)

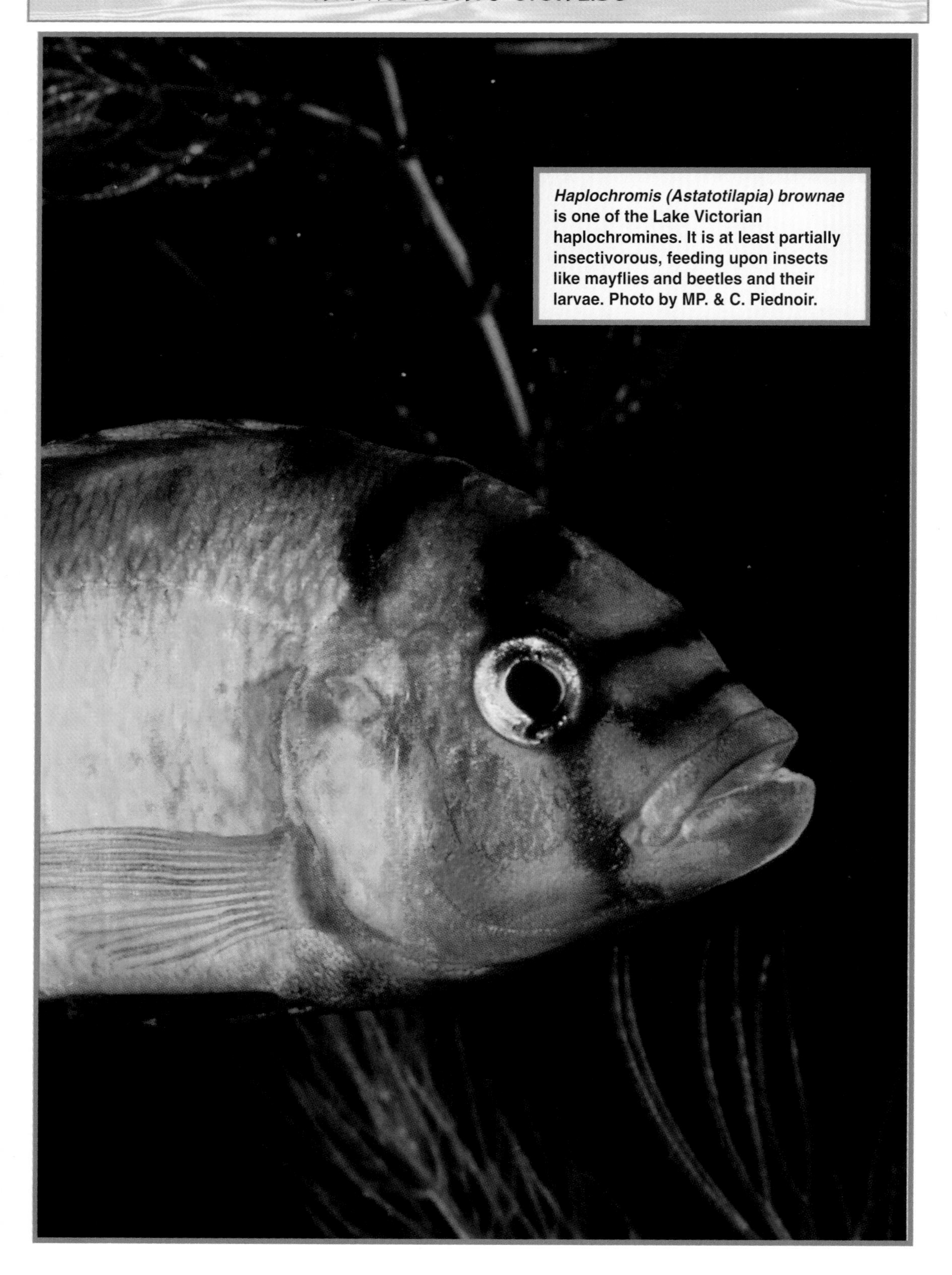

*Haplochromis (Astatotilapia) brownae* is one of the Lake Victorian haplochromines. It is at least partially insectivorous, feeding upon insects like mayflies and beetles and their larvae. Photo by MP. & C. Piednoir.

# AFRICAN CICHLIDS

Although African cichlids are synonymous with the cichlids of the rift lakes of Africa in the minds of most tropical fish hobbyists, I want to introduce the readers to some of the "other cichlids" of Africa, some of which just happen to be favorites of mine. While the cichlids of the rift lakes are well known for being adapted to hard and alkaline water, these cichlids are often from soft water areas, but many are quite adaptable.

- *Chromidotilapia guentheri*

There is no popular name for this species, but it sure has been popular with advanced cichlid hobbyists. One of the reasons is its very interesting behavior. This is a mouthbrooding cichlid with a definite difference! In most species, the female broods the eggs, and usually she is on her own for the protection of the fry. In this species, it is the male that incubates the eggs, but the female does not abandon him. She guards him. In the meantime, she is able to eat so she can build up energy reserves for the next clutch of eggs. Staying with the male and protecting him from any harassment by other fishes, she then helps guard the free-swimming fry once he releases them. In addition, this species is interesting because it is quite possible that they are naturally monogamous in the wild.

The species was once known as *Pelmatochromis guentheri* and was of particular interest because it was the only *Pelmatochromis* that was a mouthbrooder. Now we know why!

***Pelvicachromis pulcher*, commonly known as the krib, is one of the few fishes where the female is more colorful than the male. This is due to the red belly of the female, which becomes larger and redder as she fills with eggs. Photo by MP. & C. Piednoir.**

- *Hemichromis bimaculatus*
  Jewel Cichlids

Actually, there are several jewel cichlids that look quite similar, but they vary somewhat in size and intensity and detail of coloration. Paul V. Loiselle is the expert on this group, and he says that the species that has generally been known under this name in the hobby is actually *Hemichromis guttatus*. The real jewel cichlid known by this name is actually more colorful than the ones that have been in the hobby so long.

Jewel cichlids are substrate spawners, very similar in behavior to the Central American substrate spawners, quite fierce in protection of their young. In my experience, the various species are quite adaptable to a variety of water conditions. They are one of the tough cichlids, both in hardiness and formidability in defense of their young.

Although similar in appearance, the jewelfishes are fascinating to study in

*Chromidotilapia guentheri.* This is an aggressive fish, especially among its own kind. Their unique spawning habits set these fish apart from the crowd.

*Pelvicachromis subocellatus.* This lovely peaceful fish doesn't show its true colors until spawning time, when it becomes both vividly colored and mighty in defense of its family. Photo by K. Knaack.

terms of behavior, and many of them are absolutely gorgeous when mature, particularly at spawning time! Although centered in West Africa, this group is quite successful and has been found throughout the continent.

- *Pelvicachromis pulcher* Kribensis

Known throughout most of its history in the aquarium hobby as *Pelmatochromis kribensis*, the former erroneous scientific name has stuck with this species. It is one that generally does better in soft water and looks its best in that type water, too. It has been popular in the hobby because it is small, with the male only attaining a size of three inches, and it is peaceful for a cichlid; however, it is too aggressive to be kept with *Apistogramma* species of South America.

- *Pelvicachromis subocellatus*

There is no popular name for this species, but it is an absolutely gorgeous animal, especially during spawning time. The courtship is fascinating to watch. The female locates a cave and then courts a male. At this time, the red coloration in the ventral region is so intense that it seems as though it would glow in the dark. When the female courts the male, she turns her caudal fin back, displaying the bright red and violet coloration. Although this species is even a little smaller than the krib, the young take newly hatched brine shrimp once they are free swimming. They also take finely powdered dry food. In the older literature, this species may appear as "*Pelmatochromis subocellatus.*"

**THE TILAPIAS**

These fish are important because they have been too successful...with the help of proud and arrogant humankind. They were already successful in Africa, but they have been introduced as food fishes around the world, and I have seen them everywhere from Tahiti (in both fresh and salt water!) to Central America. Although some of the species are quite good looking and have long been kept in aquaria, I am not an advocate of introduction of exotic fishes to any waters. And it is quite disconcerting to be looking for some rare fish species in Central America only to find that they have seemingly been replaced by a tilapia of some kind. The original genus was *Tilapia* but Trewavas broke the genus down into three separate genera. These include *Oreochromis*, which are all mouthbrooding cichlids, with the female brooding the eggs, and *Sarotherodon*, which is intermediate between *Oreochromis* and *Tilapia*. That is, the species are still mouthbrooders, but the males and females stay together and either the male or female or both may pick up the eggs once they have been fertilized. The true *Tilapia* species are substrate spawners, with both parents tending the young.

All the different species have thrived in all localities. Some popular aquarium specimens of each type have been *Oreochromis mossambicus*, *Sarotherodon melanotheron*, and *Tilapia zillii*. Although each gets a little large, they spawn readily and have a handsome appearance. They are so

A female krib with some of her brood. To successfully raise the fry of dwarf cichlids, you must first provide microscopic foods like infusoria and later progress to newly hatched brine shrimp. Photo by MP. & C. Piednoir.

The fish in this tank are in sorry condition. When keeping aggressive fishes together, you must be alert for injuries that can lead to more severe bacterial and fungal diseases. Virtually every fish in this tank is damaged and without treatment they will die.

*Tilapia zillii*. Drawing by J. R. Quinn.

adaptable that I can find specimens in the lower reaches of the Colorado River, as well as in the Salton Sea.

### CICHLIDS OF LAKE VICTORIA

The cichlids of Lake Victoria have been of particular interest lately because they are all endangered. Not only was a predator introduced, the Nile perch, of which the fish are unafraid since they didn't evolve with it, but forest destruction and farming around the lake has made for muddy runoffs. The loss in visibility has kept the cichlids from breeding or has caused them to interbreed, with the subsequent loss of species. In any case, there is a serious species maintenance program in existence that is run by scientists and hobbyists. To many hobbyists and some scientists, the Victorian cichlids lack the interest of the others because of the young age of the lake (in geologic terms).

- *Haplochromis (Astatotilapia) brownae*

All of the cichlids of Lake Malawi and Lake Victoria are believed to have evolved from a *Haplochromis*-like ancestor. The difference is that the cichlids of Lake Malawi have had much longer to evolve. Lake Victoria is a saucer-shaped lake and, thus, is not one of the rift lakes, although it is often confused with them. The point is that Lake Victoria is a uniformly shallow lake (150 feet deep at the deepest parts), filled with lots of reeds and heavy plant growth. This species and the next are fairly typical. Although Lake Victoria has water with a high pH, it is not as hard as that of rift lakes.

*Oreochromis mossambicus*. This mouth with fins is a virtual eating machine, not ever to be confused with a community fish. Photo by MP. & C. Piednoir.

- *Haplochromis (Astatotilapia) nubilus*

Unfortunately, not many species of the Victorian cichlid flock have made it to this country. This is one that has, and it is typical of most of the species known to aquarists. (Scientists know that there were many specialized forms, but are they still there? Most are believed to be extinct. Since there were over three hundred species endemic to Lake Victoria, the loss of this group is a substantial one.) All the aquarium species are maternal mouthbrooders. They should be kept in groups. That is, there should be one male to several females. This can be a difficult situation to attain because shippers tended to send more males than females because they are the colorful ones.

- *Haplochromis obliquidens*

Like most of the other cichlids of Lake Victoria known to aquarists, this fish only reaches a length of about three inches. It should be kept in groups of one male to several females, like the others. Even the original ichthyologists who did the original descriptions raved about the colors of these fish species. Think what we have lost because of the indifference of our species!

***Haplochromis brownae.*** **There are really only three things you need to do to have success with your cichlids: keep good water quality in the aquarium, give the fish good food, and really look at your fish. If you do these things, you will seldom have real problems keeping cichlids. Photo by MP. & C. Piednoir.**

### Lake Tanganyika Cichlids

These are the cichlids that have the most variety in specializations of all the rift lakes of Africa, and they have consequently been quite popular with advanced hobbyists.

- *Boulengerochromis microlepis*

This is the so-called emperor cichlid. It is very likely the largest cichlid in the world. It builds a huge nest, and both parents protect the eggs and later the young. As adults, these fish patrol the open waters and the deeper

***Haplochromis nubilis.*** **The *Haplochromis* from Lake Victoria are distinguished by their being rather slender bodied with large egg dummies and red in the caudal fin, especially if the caudal peduncle is blackish in the breeding male. Drawing by J. R. Quinn.**

perpetual twilight areas in search of prey. The parents protect the fry for at least two weeks. When the fry are on their own, they form a strange sphere shape while schooling. Even the fry are predators, though they obviously must find fish or crustaceans smaller than themselves to prey upon.

This fish is fascinating enough that hobbyists have kept it and spawned it, but it needs a really large tank. Remember that it can easily reach a yard in length. Even though it patrols the open waters, it seems to be able to adapt to life in the aquarium, since it has spawned in captivity.

• *Callochromis macrops*

This species lives in sandy areas and feeds by sifting the sand for invertebrates. Its delicate coloration is nothing short of gorgeous, but it won't show up in really bright light. That is, it won't show up in a brightly lit tank. Sun shining on it makes it look radiant. But a tank housing this species should be kept dimly lit. The tank can be furnished with driftwood and a rock or two for decoration, but the main thing the fish wants is lots of sand to sift. Even though it is a specialized feeder, it is a good eater of dry and frozen foods, but you want foods that drop to the bottom. Eventually, the fish learn to come to the surface for foods.

The species is a maternal mouthbrooder, with the female incubating the eggs and caring for the young. The male builds a crater nest to attract females, but once she picks up the eggs, he drives her off and awaits the next female. Protection of the fry after they are released is not as intense as it is in most other cichlid species, so the fry should be raised in a tank of their own. The species can reach a maximum length of six inches.

*Callochromis macrops* has a subtle beauty to it and has an instantly recognizable cichlid attitude to it, and yet it is much different from other cichlids. Some cichlidophiles don't like it because of that reason, but others specialize in the genus, which contains three known species.

• *Chalinochromis brichardi*

Most people believe that there are many species of *Chalinochromis*, as there is much variation in members of the genus. Hobbyists tend to refer to the specimens with the lines down the body as *"bifrenatus."* The genus was only erected in 1974, so it may take a while to sort the specimens out from the very similar *Julidochromis*, *Telmatochromis*, and some *Lamprologus* species. The ones referred to as *brichardi* in the hobby are those with the beige body and the mask around the head. The specimens look benign and are slow moving, gliding along the rockwork, but they can be quite aggressive. The best way to spawn them is to get at least six young ones and raise them up. They spawn in caves or underneath rocks. They only lay a few eggs at a time normally, and you don't know that you have young until the fry start appearing from underneath the rocks. The eggs are laid on the "ceiling" of the cave, and the tighter the quarters, the better the fish seem to like

***Haplochromis obliquidens*. As far as is known, all the Victorian haps are maternal mouthbrooders. Photo by MP. & C. Piednoir.**

their cave. The young do well on dry food and the incidental food that they are able to pluck from the rocks, but they will absolutely prosper on newly hatched brine shrimp. The trouble is that you often don't know when the fish has spawned, but most hobbyists who keep this type of fish tend to feed newly hatched brine shrimp to their adults anyway. The species attains a size of about four inches, with a few rare individuals attaining a length of six inches.

- *Cyathopharynx furcifer* Furcifer

This is one of those species that may actually comprise several different species, as there is much variation according to the locality in the lake. This is one of the highly prized species of the lake, as they are quite showy fish, attaining a size of 8 inches. The males pick the highest prominence upon which to build a nest. Then they carry sand up and put on the rock. This makes them very interesting, as they are apparently descended from males of a species that used to build a nest in the sand. There were advantages to having the nest up high away from egg predators. The females tend to pick the male with the highest nest, painfully constructed from carrying all that sand up to the highest "peak" from the sand below.

***Cyathopharynx furcifer*** **is one of those fish that made Lake Tanganyika famous. There are various morphs with different basic body colors, but they are all outstanding. Photo by MP. & C. Piednoir**

One of the reasons for the popularity of this species is that males tend to be blue to purplish in coloration, and they fairly shimmer as they display before the females. The female lays her eggs in the male's nest, where he fertilizes them. She then picks them up in her mouth and broods them there and protects them for a short while after they have been released.

***Callochromis macrops.*** **These fish are similar in their feeding habits to the South American *Geophagus*. They sift through the sand for small invertebrates. Photo by MP. & C. Piednoir.**

*Chalinochromis* sp. Ndoboi. These 6-inch fish are very belligerent when crowded. Photo by M. Smith.

*Eretmodus cyanostictus.* This fish is a grazer. They like algae-covered rocks, so it's good to keep a jar of rocks in a sunny window to culture algae for them. Otherwise, offer green foods frequently. Photo by M. Smith.

The species feeds upon invertebrates in the water column, but it will also take dry foods, as well as frozen formulas. The species also thrives on newly hatched and adult brine shrimp. The fry are normally raised with newly hatched brine shrimp as a first food, but they will also take fine powdered food.

- *Cyphotilapia frontosa* Frontosa

This fish lives in the perpetual twilight of the lake at depths of about a hundred feet. It feeds upon invertebrates and small fish that it happens upon. At those depths, the supply of plankton is greater in the daytime, so some species of fish dwell there to feed upon the plankton. *C. frontosa* feeds upon the fish that feed upon the plankton, among other things. This is a very popular fish that reaches a length of nearly 14 inches. One of the reasons for its popularity, in addition to exotic appearance, is that it is an easy-going, fairly sedentary fish that does well in the aquarium. And it displays well, as it is usually just hanging there in the water near the bottom. When it does move, it does so without apparent effort.

The male constructs a rudimentary nest, and the female lays up to 20 quite large eggs. She broods these in her mouth for nearly a month, releasing young that are nearly an inch in length, looking very much like miniature adults. The young males, however, don't show the hump on the head until they reach about six inches in length. Even hobbyists who aren't fond of this fish breed it because it is always valuable commercially.

- *Eretmodus cyanostictus*

This is one of the goby cichlids, which are found in the surf zone of the rocky areas of the lake. The fish have greatly reduced swim bladders so they are able to "hop" around on the rocks and resist the swift currents of the water. They graze upon the algae and have specially adapted teeth for doing so. Although often called the "clowns of Tanganyika," these little fish (reaching about three to four inches) can be quite aggressive with one another and with other fish species, too. They need lots of aeration, as might be expected, since they live in the surf zone. Although they will take live brine shrimp avidly, it should not be fed to them as a steady diet, as it can be the death of them. They need their veggies. Hence, a diet of dry food is preferable. If you want to really make them happy, put occasional rocks out in the sun in water. Let a lot of algae grow on them and then place them back in the tank.

The species is a mouthbrooder in which both the male and female brood the eggs and young. They tend to form pairs, an unusual behavior for mouthbrooders.

- *Julidochromis ornatus*

There are several species of *Julidochromis*, including the very popular *Julidochromis marlieri* and the small *Julidochromis transcriptus*. These species are all quite similar in behavior, but those

***Julidochromis transcriptus*. Measuring in at about 3 inches, *J. transcriptus* is the smallest member of its genus. Photo by H. Hansen.**

***Julidochromis ornatus*. Photo by MP. & C. Piednoir.**

*Julidochromis dickfeldi.* This is the easiest of the "julies" to breed. Photo by MP. & C. Piednoir.

*Neolamprologus brevis* is a typical shell dweller. It is best to keep these fish with a higher ratio of females to males to prevent fighting. Photo by H. J. Richter.

who specialize in the genus would jump on me for saying that! There are subtle differences in behavior, and certainly there are differences in coloration and size. This one attains a length of about three inches. The best way to breed them is to get about six young. Eventually a pair will form and drive the others off. The eggs are jade-colored and are laid on the roof of a cave. You probably won't know the fish have spawned until young begin gliding along the rock, much like adults.

Although exceedingly quiet fish, these animals can be aggressive, but they have a certain personality to them that has made them quite popular. For one thing, they orient themselves to rocks. Wherever the rock is means "down" to them. They will glide up the side of a rock, skim along its surface, sweep down the other side, and then underneath upside down. They nearly always have a home in a spot underneath a rock. Whenever they are in their home, they will be upside down. Even when feeding at the surface of the water, they are often upside down, orienting themselves with the surface.

Another reason for the popularity of the species is that they tend to bond as a pair for life, raising only a few young at a time. Both parents defend the eggs and the young. The adults feed upon invertebrates that they find in the biocover of the rocks, but they eat nearly anything in captivity. The fry do well on powdered foods, but they would do even better on newly hatched brine shrimp. At present, there are six species of *Julidochromis*, but there are probably at least twice as many species in the lake.

- *Lamprologus brevis*

This species is a shell dweller that barely reaches two inches in length. The female is even smaller. This brave little fish survives in sandy areas where the large fish prowl. The snail shells are the secret to their success. Males hoard snail shells, as that way they get to have more females! They can have a female for each shell. The female lays eggs in the shell, and the male fertilizes them. Sometimes he can't fit in the shell, but his anal fins help guide the milt into the shell. The fish feed upon invertebrates that they find in the water and in the sand. In the aquarium, they eat anything, including dry food.

- *Lamprologus brichardi*

When this species spawns, the young help to protect succeeding siblings. The colony increases in size. The adults pick food out of the water and off of rocks. The dentition, which is quite impressive, may have evolved for defense, as well as for plucking invertebrates from rocks.

So successful has this type of fish been that it is found throughout the lake, and it is difficult to know how many species there are of this type. A complete survey of the entire lake would have to be done, and that is not likely for many years. In any case, this is a popular and elegant species. Although this fish spawns among the rocks and in caves, it normally places the eggs on the substrate, rather than upon the roof of a cave. Only a few eggs are laid at a time, but the survival rate is high.

***Lamprologus leleupi*. Leleupis are easy-to-breed cave spawners. This is a bright yellow fish that needs to be fed regularly on live foods to retain its beautiful color. Photo by H.-J. Richter**

***Lamprologus brevis* being attacked by *Julidochromis regani*. If aggression becomes a problem, rearrange your tank decorations. Photo by MP. & C. Piednoir.**

***Lamprologus calvus.*** **This fish requires live foods to thrive: brine shrimp, bloodworms, mosquito larvae, etc. Photo by MP. & C. Piednoir.**

***Neolamprologus ocellatus*** **with its shell. Photo by MP. & C. Piednoir.**

- *Lamprologus calvus*

Whereas the other *Lamprologus* species named might be called *Neolamprologus* by modern hobbyists and ichthyologists (some of them), this species would be *Altolamprologus.* It does have a very compressed shape, and it can be quite gorgeous. The shape allows it to fit into crevices in rocks and go after invertebrates and the young from other fish. The mouth makes it look Neanderthal, but it is functional for getting its prey. The species is also characterized by armor-like scales, which protect it from the attacks of small fish that may resent it when it intrudes upon their territory after invertebrates (or their fry).

Surprisingly, this compressed species likes to use snail shells for spawning, too, but it is less demanding than some of the other species that wouldn't even think about spawning without a shell. A cave will do fine. The female tends the eggs, while the male guards the area outside.

- *Lamprologus leleupi*

Leleupi

This colorful species is generally found in deep water. It only reaches a length of about two or three inches, but it is extremely aggressive. It is especially so with its own kind, but it looks for trouble with everyone. It will breed in shells, too, but it settles for caves. Most breeders keep them in a community tank of sorts for spawning, as that helps diffuse the aggression in the formation of pairs. They put about six *leleupi* in a community tank of Tanganyikan cichlids.

Ceramic caves are often used, with holes or slits just big enough for the female and male to enter. The breeder knows that there is a spawning when the female consistently stays within the cave and the male guards the outside. A standard practice is for the hobbyist to place his or her thumb over the opening and remove the cave to a rearing tank. This is usually done when the young start to stick their noses out of the cave. The female can be netted out of the rearing tank at the hobbyist's convenience.

***Lamprologus brichardi*, a handsome male with perfect fins. Photo by B. Allen.**

One of the best Tanganyikan breeders in the world, Ron Soucy, once told me to be sure to feed the young with newly hatched shrimp and color foods. If they didn't get it when they were young, they wouldn't retain the orange or yellow coloration. I tested his thesis and found that he was absolutely correct.

- *Lamprologus tretocephalus* Tret

This is one of the species for which there is always a commercial demand. It is a nice size, reaching about four inches, but will spawn at two, and it has nice coloration. The parents herd the young out in the open and give vigilant attention and defense to them. This makes the species a contrast to so many of the species that spawn in caves, shells, and under rocks. They are more like New World cichlids in that they are more obvious when they spawn. Both parents care for the fry. Of course, in a community tank, they can make life quite hard on the other species when they spawn. This is a snail eater, and I have often wondered how snails could

***Nimbochromis livingstoni*. This fish was named to honor the famous explorer Dr. David Livingstone, who is credited with being the first collector of Malawian fishes. Photo by MP. & C. Piednoir.**

*Pseudotropheus zebra*, orange variety. The zebra is widely available and easy to keep. If any fish could be credited with having started the African cichlid craze, it would have to be *Pseudotropheus zebra* in all its varieties. Photo by MP. & C. Piednoir.

*Pseudotropheus demasoni*. This is one of the dwarf *Pseudotropheus* species reaching a size of only 3.5 inches. Males are midnight blue to black with light metallic blue vertical bars and bright yellow eggspots on the anal fin. Females have the same coloration but are slightly duller. Photo by M. Smith.

even live in a lake of cichlids. Most species will include them on the menu, but the fact is that a good many of the cichlids of Lake Tanganyika are exceptions in that regard, so a homeostatic relationship has built up between the snail eaters and the snails.

- *Ophthalmotilapia boops*

When I first saw drawings of this species several decades ago, I bemusedly wondered whether it was named after Betty Boop, but the specific name refers to the eye, and it is pronounced BO-AWPS. This is a maternal mouthbrooder with an interesting breeding pattern. The males construct nests and try to entice females to them. The female lays golden eggs. When she picks them up, she mouths the ends of the male's ventral fins, which triggers the male to ejaculate sperm, fertilizing the eggs in her mouth. The gold knobs at the end of the male's fins resemble the color of the eggs. The shimmering colors of the male have made this fish an elite species among the enthusiasts of Lake Tanganyikan cichlids. There are several other species of the genus, but this is the most popular one.

- *Tropheus duboisi*

There are over a hundred known different types of *Tropheus*, but only a few of them have been described as separate species. I am only going to list two of them here as examples. This species is well known because the juveniles have a different color pattern from the adults. They are black and covered with numerous white spots, quite gorgeous in appearance. The adults have either a narrow yellow or red band on a black

***Aulonocara hansbaenschi*. This fish is a wild-caught specimen from Mutangula, Mozambique. Photo by M. Smith.**

body, depending on the geographical variation under consideration. All *Tropheus* are maternal mouthbrooders, with the females carrying the very large eggs for up to a month. The juveniles are large and look just like miniature adults. The mother protects them for a few days after release. While only a few species have been given formal scientific names, there are sure to be more to come because of all the different color morphs from the lake that do not seem to interbreed in the tank. While *Tropheus* all seem to look alike to the untrained eye, molecular analysis of the DNA of the genus revealed more differences among the different color morphs than among all of the endemic cichlids of the other lakes of Africa!

- *Tropheus moorii*

The fish under this name probably comprise several species, with many color variations that are given common or popular names, such as the "orange flame" variation, etc. Sometimes they are named according to the collecting site, such as the Kirza variety, which commonly has a broad yellow stripe at mid body.

*Tropheus* can seem impossible to keep. There are two essential elements to keeping them. There must be a minimum density of *Tropheus* species to diffuse aggression or the fish fight like crazy. This seems to be a minimum of ten individuals in the tank. They don't have to all be the same species, but they do have to be *Tropheus*. The other secret is not to feed much at all in the way of live

A young 3-inch *Tropheus moori*. Remember that this fish needs a predominantly vegetarian diet. Photo by A. Norman.

*Aulonocara korneliae*. Photo by Aaron Norman.

food. These fish live life as herbivores, but they will take the cheap calories by eating fish eggs and brine shrimp and worms, but it can be deadly to them. Especially is this true if they get a steady diet of such fare. Feed dry food that has a lot of plant matter in it. If you want to provide them with a treat, place rocks in water outside in the sun and let algae grow profusely on them. Then place one in the tank and let your *Tropheus* go crazy. Of course, *Tropheus* should have the excellent water quality and high pH that is required by all the other cichlids of Lake Tanganyika. This genus alone could provide a lifetime of study for a team of scientists.

**Comments**

This is only a sampling of the cichlids of Lake Tanganyika, and some hobbyists will be incensed that I left their favorites out. However, the intention here was to provide the reader with some idea of the variety of the lake and of the different types of species available. The vast difference in the cichlid fish species of the lake is one reason for their popularity.

## THE CICHLIDS OF LAKE MALAWI

The only drawback to these cichlids is that they are all mouthbrooders. However, there is much variation to the method employed. Besides, there is much variation in feeding behavior, and the colors and movement of the species are so dazzling that this group is probably the one most responsible for the popularity of the cichlid family with tropical fish hobbyists.

***Melanochromis auratus*. The overhanging snout of this fish is perfectly designed for grazing on the algae-covered rocks in the lake. Photo by MP. & C. Piednoir.**

As for scientists, they always have a project at the lake to study both behavior and classification, and they have found both areas of study a delight of complexity.

• *Aulonocara hansbaenschi*

This is one of the peacock cichlids. There are far too many to list, but they come in all colors, from flaming yellow to bright blue to tricolored. More quiescent than many other species in the lake, these are generally deepwater species with special sensing pores on the jaws, which enables them to "hear" crustaceans in the sand. They hover over the sand and then suddenly dive in and grab a meal.

The best way to breed these guys is to keep one male with several females, and that is the best method with most of the cichlids in this lake. With the others, the female is often just as colorful as the male, but that is not the case with all the different peacock species. But, no matter, the male has enough color for everyone!

• *Labeotropheus trewavasae*

This is one of the fish species that fueled the mania for African cichlids, as it comes in many different color forms. The most pretty (in my view) is the bright blue body with the orange dorsal fin. Both sexes display the color. A mouthbrooder like all the others, the spawning scene is rarely observed, as the male does not make a nest, and spawnings take place quickly. The female incubates a few large eggs in her mouth for several weeks and then releases young, which are miniatures of the adults. The mouth is especially adapted for scraping algae off the rocks and is located in a position that the fish can stay even with the rock while feeding.

*Melanochromis auratus*, albino. Photo by M. Smith

• *Melanochromis auratus*

This is another species that helped spark the boom in cichlid interest among tropical fish hobbyists. It was first known as *Pseudotropheus auratus* (and Lake Malawi was known as Lake Nyassa!). One of the reasons for its popularity is that even the juveniles have the bright coloration. Spawning is very similar to the foregoing species; however, the male changes coloration upon developing sexual maturity.

• *Nimbochromis livingstoni*

This species is too large for most hobbyists, but I couldn't resist listing it because of its unusual predation. It lies on its side and plays "possum," waiting for little fish to pick at the corpse. When they get close enough, the corpse comes to life, just like something out of a horror story!

• *Pseudotropheus tropheops*

This is another cichlid that fueled the desire for cichlids. It is yet one more cichlid that comes in many guises, and one of my favorite color varieties is the one in which the male is a brilliant blue and the female is a bright orange. Scientists are going crazy trying to sort out the *Pseudotropheus* genus. There are probably many more species than have been officially recognized; in fact, this one species may, in fact, represent several different species. Evolution takes place at a breathtaking rate in the lake among the cichlids. The results are a delight but very difficult to understand fully.

• *Sciaenochromis ahli*

Another fish to fire the desire of aquarists, this fish was introduced as the "electric blue." Later would come the cobalt blue. In addition, there were many other color variations of *Pseudotropheus* that bore similar names. But while *Pseudotropheus* feed upon the algae on the rocks, this species is a solitary hunter of fish. One drawback to this species is that the female is of drab coloration.

*Oreochromis mossambicus,* male. Photo by MP. & C. Piednoir.

### Comments

There is no doubt about it, the cichlids of Lake Malawi are among the most beautiful fishes in the world, and they have done a lot for the popularity of the cichlid hobby around the world. Some cichlid specialists become disdainful of these fishes because so many new people keep them. In my view, this is infantile thinking. I do find other cichlids of more interest, but I am always learning something new about these cichlids, and so are scientists. And it should be said that some of the long-term cichlid enthusiasts remain Malawi fans.